ENCYCLOPEDIA OF MATHEMATICS AND ITS APPLICATIONS

EDITED BY G.-C.ROTA

Volume 29

Combinatorial Geometries

Combinatorial Geometries

Edited by

NEIL WHITE
University of Florida

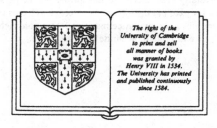

The right of the
University of Cambridge
to print and sell
all manner of books
was granted by
Henry VIII in 1534.
The University has printed
and published continuously
since 1584.

CAMBRIDGE UNIVERSITY PRESS

Cambridge

New York   Port Chester   Melbourne   Sydney

CAMBRIDGE UNIVERSITY PRESS
Cambridge, New York, Melbourne, Madrid, Cape Town, Singapore,
São Paulo, Delhi, Dubai, Tokyo

Cambridge University Press
The Edinburgh Building, Cambridge CB2 8RU, UK

Published in the United States of America by Cambridge University Press, New York

www.cambridge.org
Information on this title: www.cambridge.org/9780521070362

First published 1987
Reprinted 1990
This digitally printed version 2009

A catalogue record for this publication is available from the British Library

Library of Congress Cataloguing in Publication data
Combinatorial geometries.
(Encyclopedia of mathematics and its applications; v. 29)
Includes bibliographies and index.
1. Matroids.   2. Combinatorial geometry.   I. White,
Neil.   II. Series.
QA166.6.C66   1987   516'.13   86–31001

ISBN 978-0-521-3339-9 Hardback
ISBN 978-0-521-07036-2 Paperback

CONTENTS

CONTRIBUTORS

Martin Aigner
II Mathematisches Institut
Freie Universitat Berlin
Konigin Luise Strasse 24–26 D-1000
Berlin 33 Germany

Richard Brualdi
Department of Mathematics
University of Wisconsin
Madison, WI 53706

Raul Cordovil
C.F.M.C.
Av. Prof. Gama Pinto, 2
1699 Lisboa-Codex,
Portugal

Ulrich Faigle
Institut Für Ökonometrie
und Operations Research
Universität Bonn
Nassestrasse 2
D-5300 Bonn 1,
Federal Republic of Germany

J.C. Fournier
20, rue Jean Lurcat
F-91230 Montgeron, France

Bernt Lindström
Matematiska Institutionen
Stockholms Universitet
Box 6701
S-113 85 Stockholm
Sweden

Neil L. White
Department of Mathematics
University of Florida
Gainesville, FL 32611

Thomas Zaslavsky
Department of Mathematics
SUNY at Binghamton
Binghamton, NY 13901

SERIES EDITOR'S STATEMENT

A large body of mathematics consists of facts that can be presented and described much like any other natural phenomenon. These facts, at times explicitly brought out as theorems, at other times concealed within a proof, make up most of the applications of mathematics, and are the most likely to survive change of style and of interest.

This ENCYCLOPEDIA will attempt to present the factual body of all mathematics. Clarity of exposition, accessibility to the nonspecialist, and a thorough bibliography are required of each author. Volumes will appear in no particular order, but will be organized into sections, each one comprising a recognizable branch of present-day mathematics. Numbers of volumes and sections will be reconsidered as times and needs change.

It is hoped that this enterprise will make mathematics more widely used where it is needed, and more accessible in fields in which it can be applied but where it has not yet penetrated because of insufficient information.

Gian-Carlo Rota

PREFACE

This book is the second in a three-volume series, the first of which is *Theory of Matroids*, and the third of which will be called *Combinatorial Geometries: Advanced Theory*. The three volumes together will constitute a fairly complete survey of the current knowledge of matroids and their closely related cousins, combinatorial geometries. As in the first volume, clear exposition of our subject has been one of our main goals, so that the series will be useful not only as a reference for specialists, but also as a textbook for graduate students and a first introduction to the subject for all who are interested in using matroid theory in their work.

This volume begins with three chapters on coordinatization or vector representation, by Fournier and White. They include a general chapter on 'Coordinatizations,' and two chapters on the important special cases of 'Binary Matroids' and 'Unimodular Matroids' (also known as regular matroids). These are followed by two chapters by Brualdi, titled 'Introduction to Matching Theory' and 'Transversal Matroids,' and a chapter on 'Simplicial Matroids' by Cordovil and Lindström. These six chapters, together with Oxley's 'Graphs and Series-Parallel Networks' from the first volume, constitute a survey of the major special types of matroids, namely, graphic matroids, vector matroids, transversal matroids, and simplicial matroids. We follow with two chapters on the important matroids invariants, 'The Möbius Function and the Characteristic Polynomial' by Zaslavsky and 'Whitney Numbers' by Aigner. We conclude with a chapter on the aspect of matroid theory that is primarily responsible for an explosion of interest in the subject in recent years, 'Matroids in Combinatorial Optimization' by Faigle.

My deepest thanks are due to the contributors to this volume, and to all others who have helped, including chapter referees. I am particularly indebted to Henry Crapo for continued support in securing the graphics work for all three of these volumes. Richard Brualdi thanks the National Science Foundation for their partial support of his work under grant DMS-8320189.

University of Florida   Neil L. White

1

Coordinatizations

NEIL WHITE

1.1. Introduction and Basic Definitions

The purpose of this chapter is to provide background and general results concerning coordinatizations, while the more specialized subtopics of binary and unimodular matroids are covered in later chapters. The first section of this chapter is devoted to definitions and notational conventions. The second section concerns linear and projective equivalence of coordinatizations. Although they are not usually explicitly considered in other expositions of matroid coordinatization, these equivalence relations are very useful in working with examples of coordinatizations, as well as theoretically useful as in Proposition 1.2.5. Section 1.3 involves the preservation of coordinatizability under certain standard matroid operations, including duality and minors. The next section presents some well-known counterexamples, and Section 1.5 considers characterizations of coordinatizability, especially characterizations by excluded minors. The final five sections are somewhat more technical in nature, and may be omitted by the reader who desires only an introductory survey. Section 1.6 concerns the bracket conditions, another general characterization of coordinatizability. Section 1.7 presents techniques for construction of a matroid requiring a root of any prescribed polynomial in a field over which we wish to coordinatize it. These techniques are extremely useful in the construction of examples and counterexamples, yet are not readily available in other works, except Greene (1971). The last three sections concern characteristic sets, the use of transcendentals in coordinatizations, and algebraic representation (i.e., modeling matroid dependence by algebraic dependence). Some additional topics which could have been considered here, such as chain groups, are omitted because they are well-covered in other readily available sources, such as Welsh (1976).

Since the prototypical example of a matroid is an arbitrary subset of a finite dimensional vector space, that is, a vector matroid, and since many matroid operations have analogs for vector spaces, which are algebraic and therefore easier to employ, a natural and important problem is to determine which matroids are isomorphic to vector matroids. This leads directly to the concept of coordinatization. In this chapter we assume that matroids are finite.

A *coordinatization* of a matroid $M(S)$ in a vector space V is a mapping $\zeta: S \to V$ such that for any $A \subseteq S$, A is independent in $M \Leftrightarrow \zeta|_A$ is injective (one-to-one) and $\zeta(A)$ is linearly independent in V.

Thus we note that a dependent set in M may either be mapped to a linearly dependent set in V or mapped non-injectively.

We note that $\zeta(s) = 0$ if and only if s is a loop. Moreover for non-loops s and t, $\zeta(s)$ is a non-zero scalar multiple of $\zeta(t)$ if and only if $\{s, t\}$ is a circuit (i.e., s and t are parallel). Thus $\zeta(s) = \zeta(t)$ only if $\{s, t\}$ is a circuit, and we see that non-injective coordinatizations exist only for matroids which are not combinatorial geometries. Furthermore, we also see that coordinatizing a matroid is essentially equivalent to coordinatizing its associated combinatorial geometry.

If B is any basis of $M(S)$, then let W be the span of $\zeta(B)$ in V. Then dim $W = $ rk M and $\zeta(S) \subseteq W$. Thus we may restrict the range of ζ to W, and thus, without loss of generality, all coordinatizations will be assumed to be in a vector space of dimension equal to the rank of the matroid. If n is the rank of $M(S)$, then for a given field K there is, up to isomorphism, a unique vector space V of dimension n over K. Thus we may also speak of a *coordinatization of M over K*, meaning a coordinatization in V.

Let $GF(q)$ denote the finite field of order q. A matroid which has a coordinatization over $GF(2)$, or $GF(3)$, is called *binary*, or *ternary*, respectively. A matroid which may be coordinatized over every field is called *unimodular* (or *regular*). Further characterizations of these classes of matroids will be given later in this chapter and in the following chapters.

It is often convenient to represent a coordinatization in matrix form. If $\zeta: S \to V$ is a coordinatization of $M(S)$ of rank n, and E a basis of V, let $A_{\zeta,E}$ be the matrix with n rows and with columns indexed by S whose a-th column, for $a \in S$, is the vector $\zeta(a)$ represented with respect to E. Since the matrix $A_{\zeta,E}$ also determines the coordinatization ζ if we are given E, we often simply say $A_{\zeta,E}$ is a coordinatization of $M(S)$.

1.2. Equivalence of Coordinatizations and Canonical Forms

If $\phi: V \to V$ is a non-singular linear transformation and $\zeta: S \to V$ is a coordinatization of $M(S)$, then $\phi \circ \zeta: S \to V$ is also a coordinatization. If Q is the non-singular $n \times n$ matrix representing ϕ with respect to the basis E of V, then $A_{\phi \circ \zeta, E} = Q A_{\zeta, E}$. On the other hand, we may easily check that

$A_{\phi \circ \zeta, E} = A_{\zeta, \phi^{-1}E}$, so multiplying $A_{\zeta, E}$ on the left by Q may also be regarded as simply a change of basis for the coordinatization ζ.

We recall from elementary linear algebra that multiplying $A_{\zeta, E}$ on the left by a non-singular matrix Q is equivalent to performing a sequence of elementary row operations on $A_{\zeta, E}$, and that any such sequence of elementary row operations on $A_{\zeta, E}$ may be realized by an appropriate choice of Q. We will say $A_{\zeta, E}$ and $QA_{\zeta, E}$ are *linearly equivalent* (where Q is non-singular), and any matrix linearly equivalent to $A_{\zeta, E}$ may be regarded as representing the same coordinatization ζ of the same matroid with respect to a new basis of V.

Conversely, given a coordinatization matrix $A_{\zeta, E}$, we may choose any new basis E' of V, and $A_{\zeta, E'}$ is linearly equivalent to $A_{\zeta, E}$. As a special case of this, we pick $E' = \zeta(B)$, where B is a fixed basis of the matroid $M(S)$.

Then, by reordering the elements of S so that the first n elements are the elements of B, we have a matrix $A_{\zeta, E'}$ in echelon form

$$B \quad S - B$$
$$A_{\zeta, E'} = \begin{pmatrix} I_n | & L \end{pmatrix}$$

where I_n is the $n \times n$ identity matrix, with columns indexed by B, and L is an $n \times (N - n)$ matrix with columns indexed by $S - B$, where $N = |S|$.

As yet another way of viewing linear equivalence, let W_ζ be the subspace spanned by the rows of $A_{\zeta, E'}$ in an N-dimensional vector space U. What we have seen is that W_ζ is independent of E', and that indeed the choice of E' actually amounts to a choice of a basis for W_ζ. Thus every linear equivalence class of $n \times N$ matrices coordinatizing $M(S)$ corresponds to an n-dimensional subspace of U. Conversely, every n-dimensional subspace of U corresponds to a coordinatization of some rank n matroid on S, which is a weak-map image of $M(S)$.

Remark. Algebraic geometers regard the collection of all n-dimensional subspaces of an N-dimensional vector space as a Grassmann manifold, and the coordinatizations of $M(S)$ correspond to a certain submanifold.

Besides row operations, another operation on $A_{\zeta, E}$ which leaves invariant the matroid coordinatized by $A_{\zeta, E}$ is non-zero scalar multiplication of columns. This may be accomplished by multiplying $A_{\zeta, E}$ on the right by an $N \times N$ diagonal matrix with non-zero diagonal entries. Combining this with the previous operations, we say that two $n \times N$ matrices A and A' are *projectively equivalent* if there exist Q, an $n \times n$ non-singular matrix, and D, an $N \times N$ non-singular diagonal matrix, such that $A' = QAD$.

Let us recall that projective $n - 1$ dimensional space P is obtained from V by identifying the non-zero vectors of each one-dimensional subspace of V to give a point of P. Let $\pi : V \rightarrow P \cup \{0\}$ be the resulting map, where 0 is an element adjoined to P which is the image of $0 \in V$. Then if $\zeta : S \rightarrow V$ is a coordinatization,

$\pi \circ \zeta$ is an embedding of $M(S)$ into $P \cup \{0\}$, except that parallel elements become identified in $P \cup \{0\}$. If $T': V \to V$ is a linear transformation, let $T = \pi \circ T' \circ \pi^{-1}$, which is well-defined since T' preserves scalar multiples. Then we call T a linear transformation of $P \cup \{0\}$. Since non-zero scalar multiples in V are identified in $P \cup \{0\}$, we immediately have the following:

1.2.1. Proposition. *Let J and L be $n \times N$ matrices over the field K. Then if J coordinatizes $M(S)$ and J is projectively equivalent to L, then L also coordinatizes $M(S)$. J and L are projectively equivalent if and only if their corresponding coordinatizations ζ_J and ζ_L determine the same projective embedding up to change of basis in $P \cup \{0\}$, i.e., $\pi \circ \zeta_J = T \circ \pi \circ \zeta_L$, where T is a non-singular linear transformation of $P \cup \{0\}$.*

We next ask whether there exists a canonical form for a projective equivalence class of coordinatizations, as echelon form was for a linear equivalence class. For a given coordinatization

$$A = (I_n | L)$$

in echelon form with respect to a basis B, let L^+ be the matrix obtained by replacing each non-zero entry of L by 1. In fact, L^+ is just the incidence matrix of the elements of B with the basic circuits of the elements of $S - B$, so it is independent of the particular coordinatization. Now let Γ be the bipartite graph whose adjacency matrix is L^+. Thus each entry of 1 in L^+ corresponds to an edge of Γ. Let T be a basis (i.e., spanning tree) of Γ.

1.2.2. Proposition. *(Brylawski and Lucas, 1973) A is projectively equivalent to a matrix A' which is in echelon form with respect to B, and which has 1 for each entry corresponding to an edge of T.*

Proof. This may be accomplished by non-zero scalar multiplication of rows and columns, and is left as an exercise. $\qquad\square$

The matrix A' of the preceding proposition is said to be in (B, T)-*canonical form*, or when B and T are understood, *canonical projective form*. The simplest canonical projective form and most useful version of this canonical form occurs when $M(S)$ has a spanning circuit C. Then by choosing B to be $C - \{c\}$ for some $c \in C$, the column corresponding to c in L has no zeros, hence we may pick T to correspond to the n entries of column c, together with the first non-zero entry in every other column of L.

A major use of this projective canonical form is in actual computation with coordinates and in presenting examples.

1.2.3. Example. Let $M(S)$ be the 8-point rank 3 geometry whose affine diagram appears in Figure 1.1. If we choose the standard basis $B = \{b_1, b_2, b_3\}$

Figure 1.1. An 8-point rank 3 geometry.

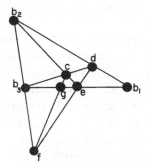

and spanning circuit $C = \{b_1, b_2, b_3, c\}$, we may coordinatize M over $\mathbb{Q}$ by the following matrix in canonical projective form:

$$
\begin{array}{cccccccc}
b_1 & b_2 & b_3 & c & d & e & f & g \\
\end{array}
$$
$$
\begin{pmatrix}
1 & 0 & 0 & 1 & 1 & 1 & 0 & 1 \\
0 & 1 & 0 & 1 & 1 & 0 & 1 & 0 \\
0 & 0 & 1 & 1 & 0 & 1 & -1 & 2
\end{pmatrix}.
$$

1.2.4. Example. Let $M(S)$ be the 4-point line, that is, $U_{2,4}$, the uniform geometry of cardinality 4 and rank 2, whose bases are all of the subsets of S of cardinality 2, where $|S| = 4$. Then any coordinatization of $M(S)$ over any field K may be put in the following projective echelon form:

$$
\begin{array}{cccc}
1 & 0 & 1 & 1 \\
0 & 1 & 1 & \alpha
\end{array}
$$

where $\alpha \in K - \{0, 1\}$. Thus we can say that up to projective equivalence, there is a one-parameter family of coordinatizations of $U_{2,4}$. We note that this parameter α is equivalent to the classical cross-ratio of four collinear points in projective geometry.

Since $U_{2,4}$ is the simplest non-binary matroid, one might be led to surmise the following, first proved by White (1971, Proposition 5.2.5), and later by Brylawski and Lucas (1973) using more elementary techniques. The proof is omitted here, because of its fairly technical nature.

1.2.5. Proposition. *Let $M(S)$ be a binary matroid and K a field over which M has a coordinatization. Then any two coordinatizations of M over K are projectively equivalent.*

Brylawski and Lucas (1973) have investigated t' ₂ question of which matroids have, over a particular field K, any two coordinatizations projectively equivalent. Such matroids are said to be *uniquely coordinatizable over K*, and among their findings is that ternary matroids are uniquely coordinatizable over $GF(3)$ (although not over an arbitrary field, as the example of $U_{4,2}$ shows).

1.2.6. Example. We return to Example 1.2.3. This example is, in fact, a ternary matroid, which is uniquely coordinatizable not only over $GF(3)$, but over every field K such that char $K \neq 2$. To see this, we first note that the matrix given over $\mathbb{Q}$ may be regarded as a coordinatization of M over every field K such that char $K \neq 2$. If we take an arbitrary coordinatization of M over any such field K and put that coordinatization in canonical projective form with respect to B and C, the elements b_1, b_2, b_3, and c are assigned the vectors shown, and then the vector for d is determined since d is on the intersection of the two lines $b_1 b_2$ and $b_3 c$. Likewise $e \in b_1 b_3 \cap b_2 c$, $f \in b_2 b_3 \cap de$, and $g \in b_1 b_3 \cap cf$.

1.3. Matroid Operations

We now note that coordinatizability is preserved under various matroid operations, including duality, minors, direct sums, and, in a restricted sense, truncation. This material is also found scattered through Chapter 7 of White (1986), and is collected here for convenience.

1.3.1. Proposition. *Let $A_{\zeta,E}$ coordinatize $M(S)$ over a field K, and let W_ζ be the row-space of $A_{\zeta,E}$ in U, a vector space of dimension $N = |S|$ over K. Then if $M^*(S)$ denotes the dual matroid of M, the subspace $W_\zeta^\perp$ orthogonal to W_ζ is the subspace of U corresponding to a coordinatization of M^*. Thus M is coordinatizable over K if and only if M^* is.*

Furthermore, if $A_{\zeta,E}$ is in echelon form, $A_{\zeta,E} = (I_n, L)$, then $A^* = (-L^t, I_{N-n})$ is a coordinatization of M^*, where t denotes transpose.

Proof. Let B be a basis of $M(S)$ and we may assume $A_{\zeta,E}$ is in echelon form with respect to B, since W_ζ is invariant under linear equivalence. Thus $A_{\zeta,E} = (I_n, L)$, and we note that $A^* = (-L^t, I_{N-n})$ has each of its rows orthogonal to each row of $A_{\zeta,E}$, hence the rows of A^* are a basis of $W_\zeta^\perp$. Let $M'(S)$ be the matroid coordinatized by the columns of A^*. Since $S - B$ corresponds to the columns

Figure 1.2. A 7-point rank 3 matroid M.

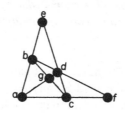

of I_{N-n} in A^*, we see that $S - B$ is a basis of M'. Conversely, if B' is any basis of M', $S - B'$ is a basis of M by a similar argument. Since B was an arbitrary basis of M, $M' = M^*$ and the theorem follows.   □

1.3.2. Example. Let $M(S)$ be the 7-point rank 3 matroid shown in Figure 1.2, along with a coordinatization A over $\mathbb{R}$ given below. Then M^*, a rank 4 matroid which is shown in Figure 1.3, has the coordinatization A^* over $\mathbb{R}$ as in the preceding proposition.

$$
A = \begin{array}{c} \begin{array}{ccccccc} a & b & c & d & e & f & g \end{array} \\ \begin{pmatrix} 1 & 0 & 0 & 1 & 1 & 1 & 0 \\ 0 & 1 & 0 & 1 & 1 & 0 & 1 \\ 0 & 0 & 1 & 1 & 0 & 1 & 1 \end{pmatrix} \end{array},
$$

$$
A^* = \begin{array}{c} \begin{array}{ccccccc} a & b & c & d & e & f & g \end{array} \\ \begin{pmatrix} -1 & -1 & -1 & 1 & 0 & 0 & 0 \\ -1 & -1 & 0 & 0 & 1 & 0 & 0 \\ -1 & 0 & -1 & 0 & 0 & 1 & 0 \\ 0 & -1 & -1 & 0 & 0 & 0 & 1 \end{pmatrix} \end{array}
$$

Figure 1.3. M^*, the dual of the matroid M in Figure 1.2, where $abfg$, $aceg$, $bcef$ are coplanar sets.

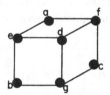

1.3.3. Proposition. *Let $M(S)$ be a matroid.*
  (1) *If M is coordinatizable over a field K, then so is every minor of M.*
  (2) *If $M = M_1 \oplus M_2$, then M is coordinatizable over K if and only if both M_1 and M_2 are coordinatizable over K.*
  (3) *If K is sufficiently large and M is coordinatizable over K, then the truncation $T(M)$ is coordinatizable over K.*

Proof. (1) If $A_{\zeta,E}$ coordinatizes M, then any submatroid $M - X$ is coordinatized by deleting the columns of $A_{\zeta,E}$ corresponding to X. Since contraction is the dual operation to deletion, (1) follows from the preceding proposition. For a direct construction of a coordinatization of a contraction, see the following remark and example.
  (2) If $A^{(1)}$ and $A^{(2)}$ are matrices coordinatizing M_1 and M_2 respectively, then

the matrix direct sum

$$\begin{pmatrix} A^{(1)} & 0 \\ 0 & A^{(2)} \end{pmatrix}$$

is a coordinatization of $M = M_1 \oplus M_2$. The converse follows from (1).

(3) The construction of truncation (to rank $n-1$, say) described in Section 7.4 of White (1986) may be carried out within the vector space V provided only that the field is sufficiently large to guarantee the existence of a free extension (by one point) within V.   □

1.3.4. Remark. To construct the coordinatization of a contraction $M(S)/X$ from a coordinatization $A_{\zeta,E}$ of M, we first choose a basis I of the set X. By row operations on $A_{\zeta,E}$ we may make the first $n-k$ entries 0 in each column corresponding to I, where $k = |I|$. Then delete the columns corresponding to X, as well as the last k rows.

This construction really amounts to simply taking a linear transformation T from V, the vector space in which M is coordinatized, to a vector space of dimension $n-k$, such that the kernel of T is precisely span (ζX).

1.3.5. Example. Let M be the matroid shown in Figure 1.4, with coordinatization A over $\mathbb{Q}$. Let $X = \{e, f\}$. Then row operations on A lead to the matrix A', and deletion of the appropriate rows and columns gives A'', a coordinatization of M/X, which is put into canonical projective form A'''. The matroid M/X is shown in Figure 1.5.

$$A = \begin{array}{c} \\ \\ \\ \\ \end{array} \begin{array}{cccccccc} a & b & c & d & e & f & g & h \\ \end{array}$$

$$A = \begin{pmatrix} 1 & 0 & 0 & 0 & 1 & 1 & 1 & 0 \\ 0 & 1 & 0 & 0 & 3 & 3 & 0 & 0 \\ 0 & 0 & 1 & 0 & -2 & 7 & 0 & 1 \\ 0 & 0 & 0 & 1 & 0 & 0 & 2 & -5 \end{pmatrix},$$

$$\begin{array}{cccccccc} a & b & c & d & e & f & g & h \\ \end{array}$$

$$A' = \begin{pmatrix} 0 & 0 & 0 & 1 & 0 & 0 & 2 & -5 \\ -3 & 1 & 0 & 0 & 0 & 0 & -3 & 0 \\ 0 & 0 & 1 & 0 & -2 & 7 & 0 & 1 \\ 1 & 0 & 0 & 0 & 1 & 1 & 1 & 0 \end{pmatrix},$$

$$\begin{array}{cccccc} a & b & c & d & g & h \\ \end{array}$$

$$A'' = \begin{pmatrix} 0 & 0 & 0 & 1 & 2 & -5 \\ -3 & 1 & 0 & 0 & -3 & 0 \end{pmatrix},$$

$$\begin{array}{cccccc} d & b & g & a & c & h \\ \end{array}$$

$$A''' = \begin{pmatrix} 1 & 0 & 1 & 0 & 0 & 1 \\ 0 & 1 & 1 & 1 & 0 & 0 \end{pmatrix}.$$

Figure 1.4. A matroid M.

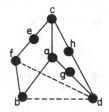

Figure 1.5. M/X, with M as Figure 1.4.

1.4. Non-coordinatizable Geometries

We now give several examples of combinatorial geometries which may not be coordinatized over any field.

The first example is a rank 3 matroid obtained from the Desargues configuration by replacing the 3-point line klm by three 2-point lines, kl, km, and lm, as shown in Figure 1.6. Coordinatization of this matroid over a field K is equivalent to embedding this configuration in the projective plane $P(2, K)$. However, $P(2, K)$ is a Desarguesian plane, which means simply that in this configuration, klm must be collinear, so coordinatization is impossible. This matroid is called the non-Desargues matroid.

Figure 1.6. The non-Desargues matroid.

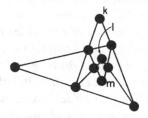

A second example of a non-coordinatizable geometry, the non-Pappus matroid, is obtained from the Pappus configuration in a manner similar to that just given for the Desargues configuration. This is illustrated in Figure 1.7, where x, y, and z are non-collinear, violating the usual assertion of Pappus' Theorem.

A third example is a class of examples which are the smallest non-coordinatizable geometries in terms of cardinality. The simplest member of this class, discovered by Vámos (1971), is described by letting $S = \{a, b, c, d, a', b', c', d'\}$, and letting the bases of $M(S)$ be all the 4-element

Figure 1.7. The non-Pappus matroid.

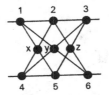

Figure 1.8. A Vámos cube.

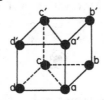

subsets of S except $aa'\,bb'$, $bb'\,cc'$, $cc'\,dd'$, $aa'\,dd'$, $aa'\,cc'$. This matroid of rank 4 may be illustrated by the affine diagram in Figure 1.8, even though it cannot actually exist in an affine space as a consequence of its non-coordinatizability.

First we verify that M is actually a combinatorial geometry. This is easy in terms of circuit exchange. The circuits of M are the five 4-element subsets which are not bases, as listed above, together with each 5-element subset of S which does not contain any of the 4-element circuits. Now if C_1 and C_2 are circuits with $C_1 \neq C_2$, and $x \in C_1 \cap C_2$, we first note that $|C_1 \cup C_2| \geq 6$, since circuits are incomparable and no two of the 4-element circuits have an intersection of more than two elements. Hence $(C_1 \cup C_2) - x$ has cardinality at least 5, and contains a circuit. Hence $M(S)$ is a geometry.

Next we show that M is, in fact, non-coordinatizable. Suppose, to the contrary, that M has been embedded in $P(3, K)$ for some field K. Then dd', which is not coplanar with $aa'cc'$, must intersect the plane $aa'cc'$ in a point e. But since $e \in aa'dd' \cap cc'dd'$, we must have $e \in aa' \cap cc'$. By a symmetric argument, bb' must also intersect $aa'cc'$ in e, but then b, b', d, and d' are coplanar, contradicting the fact that $bb'dd'$ is a basis of $M(S)$.

Finally we note that further members of this class of examples may be constructed by taking the same set S and the five 4-element circuits given for M, and then listing additional 4-element circuits (and letting all other 4-element subsets of S remain as bases) subject to two constraints:

(i) $bb'dd'$ remains a basis;

(ii) no two of the 4-element circuits intersect in more than two elements. The argument that the result is a combinatorial geometry which is non-coordinatizable proceeds exactly as above.

The member of this class of examples which has the maximum number of 4-element circuits is the one which has, besides the five given 4-element circuits,

$abcd, a'b'c'd', abc'd', ab'cd', ab'c'd, a'bcd', a'bc'd$, and $a'b'cd$. If $bb'dd'$ were also to be made a circuit, the resulting geometry would be isomorphic to a three-dimensional binary affine space, $AG(3, 2)$.

The members of a fourth (and very large) class of non-coordinatizable geometries are obtained by taking two geometries G_1 and G_2 such that there is no field over which both G_1 and G_2 may be coordinatized, and then constructing a geometry G_3 which has both G_1 and G_2 as minors. There are many ways of constructing such a geometry G_3, with perhaps the two most natural being the direct sum of G_1 and G_2, and the direct sum truncated to a rank equal to the rank of G_1 or G_2, whichever is larger.

1.5 Necessary and Sufficient Conditions for Coordinatization

The most successful coordinatization conditions are the excluded minor characterizations of the classes of matroids coordinatizable over certain fields. We will discuss these first, and follow with a consideration of conditions for coordinatizability over arbitrary fields.

If A is a class of matroids, an *excluded minor characterization* of A is collection E of matroids with the property that for every matroid M, $M \in A$ if and only if there does not exist $N \in E$ with N isomorphic to a minor of M. Although E could be either finite or infinite, we are primarily interested in this type of characterization when E is finite. It is elementary to check that A has an excluded minor characterization if and only if A is a *hereditary class*, that is, a class of matroids closed under the taking of minors.

The class of binary matroids is by far the best understood class of matroids, because of its particularly simple structure.

1.5.1. Proposition. *A matroid is binary if and only if it has no minor isomorphic to the 4-point line, $U_{2,4}$.*

This and many other characterizations of binary matroids are given in Chapter 2.

A particular binary matroid we will frequently refer to is F_7, the Fano plane, given by the following binary coordinatization:

$$\begin{array}{cccccccc} & a & b & c & d & e & f & g \\ \begin{pmatrix} 1 & 0 & 0 & 1 & 1 & 0 & 1 \\ 0 & 1 & 0 & 1 & 0 & 1 & 1 \\ 0 & 0 & 1 & 0 & 1 & 1 & 1 \end{pmatrix} \end{array}.$$

This matroid is also sometimes referred to as $PG(2, 2)$, the projective plane over $GF(2)$, and is illustrated in Figure 1.9.

Figure 1.9. The Fano matroid, F_7.

The excluded minor characterization of ternary matroids was discovered and proved by R. Reid, *c.* 1971, but never published. The result, which follows, was published independently by Bixby (1979) and by Seymour (1979).

1.5.2. Proposition. *A matroid is ternary if and only if it has no minor isomorphic to one of*

$$U_{2,5}, U_{2,5}^* \ (\text{which is } U_{3,5}), F_7, \text{ or } F_7^*.$$

A third excluded minor characterization, that of unimodular matroids by Tutte (1958), stands as one of the crowning achievements of matroid theory. This theorem is very deep, as it was first proved by way of Tutte's Homotopy Theorem. There are other proofs now available which are more elementary (Seymour 1979).

1.5.3. Theorem. *A matroid is unimodular if and only if it has no minor isomorphic to one of*

$$U_{2,4}, F_7, F_7^*.$$

Another equally striking characterization of unimodular matroids was found by Seymour (1980). He shows that every unimodular matroid may be built up in certain ways from graphic matroids, cographic matroids, and copies of a particular matroid called R_{10}.

These and several other characterizations of unimodular matroids are discussed in Chapter 3.

There are some very interesting excluded minor characterizations for several classes of graphic matroids. These characterizations are discussed more completely in Chapter 2, but are included here for the sake of completeness.

1.5.4. Theorem. (*Tutte 1959*). *A matroid is graphic if and only if it has no minor isomorphic to*

$$U_{2,4}, F_7, F_7^*, M(K_5)^*, \text{ or } M(K_{3,3})^*.$$

Here K_5 and $K_{3,3}$ are the Kuratowski graphs, the complete graph on five vertices, and the complete bipartite graph on two sets of three vertices, respectively. Also, $M(G)$ is the polygon or cycle matroid of the graph G, and $M(G)^*$ is the orthogonal matroid of $M(G)$, namely the bond matroid of G. By duality, a matroid is cographic if and only if it has no minor isomorphic to $U_{4,2}$, F_7, F_7^*, $M(K_5)$, or $M(K_{3,3})$. The excluded minor characterization of planar graphic matroids is a very pleasing generalization of Kuratowski's Theorem, which states that a graph is planar if and only if it has no homeomorphic image of a subgraph isomorphic to K_5 or $K_{3,3}$. A matroid is planar graphic if and only if it has no minor isomorphic to $U_{2,4}$, F_7, F_7^*, $M(K_5)$, $M(K_5)^*$, $M(K_{3,3})$, $M(K_{3,3})^*$, or, equivalently, if and only if it is graphic with no minor isomorphic to $M(K_5)$ or $M(K_{3,3})$. Thus the planar graphic matroids are precisely those matroids which are both graphic and cographic. One more interesting subclass of the graphic matroids is the class of series-parallel matroids, which are characterized by the excluded minors $U_{2,4}$ and $M(K_4)$.

A number of interesting relations may be deduced from these excluded minor characterizations. For example, a hereditary class is closed under duality if and only if the dual of each excluded minor is also an excluded minor. This is the case for each of the classes considered above, except graphic and cographic matroids, which are duals of each other.

We can also see that a hereditary class A is contained in another hereditary class A' if and only if every excluded minor of A' has itself some minor which is an excluded minor for A. For example, graphic and cographic matroids are unimodular, and unimodular matroids are binary as well as ternary.

We now turn to general necessary and sufficient conditions for coordinatization. The following result of Tutte was the first such set of conditions and it was also an important step in his proof of the excluded minor characterization of unimodular matroids.

1.5.5. Proposition. *Let $M(S)$ be a matroid and assume that for every hyperplane (or copoint) H of M is given a function $F_H : S \to K$, where K is a field, so that*
  (1) *kernel $F_H = H$ for every hyperplane H.*
  (2) *For every three hyperplanes H_1, H_2, H_3 of M containing a common coline, there exist constants $\alpha_1, \alpha_2, \alpha_3, \in K$, all non-zero, such that $\alpha_1 F_{H_1} + \alpha_2 F_{H_2} + \alpha_3 F_{H_3} = 0$.*
Then M may be coordinatized over K. Conversely, any coordinatization of M over K may be used to construct functions F_H satisfying (1) and (2).

In order to prove this proposition, we first need a lemma. Let W denote the vector space of all functions from S into K, and V the subspace of W spanned by $\{f_H | H \text{ is a hyperplane of } M\}$.

1.5.6. Lemma. *Let* $\{f_H\}$ *be given satisfying the hypotheses of Proposition 1.5.5, and let* $B = \{b_1, b_2, \ldots, b_n\}$ *be a basis of* $M(S)$. *Then the functions* f_{H_i} *corresponding to the basic hyperplanes* $H_i = \overline{B - b_i}$ *form a basis of* V.

Proof of lemma. $A = \{f_{H_1}, f_{H_2}, \ldots, f_{H_n}\}$ is linearly independent in V, for $f_{H_i}(b_j) \neq 0$ if and only if $i = j$, for $1 \leqslant i \leqslant n, 1 \leqslant j \leqslant n$. It remains to be shown that $f_H \in \text{span } A$ for every hyperplane H.

Let H be an arbitrary hyperplane of M, and let $h = n - 1 - |H \cap B|$. We use induction on h, noting that the case $h = 0$ is trivial, since then $f_H \in A$.

Assume by induction hypothesis that $f_J \in \text{span } A$ for all hyperplanes J such that $n - 1 - |J \cap B| < h$.

Now we assume by re-indexing that $H \cap B = \{b_1, b_2, \ldots, b_l\}, l = n - h - 1$. Since $H \cap B$ is independent, we may extend it to a basis $\{b_1, b_2, \ldots, b_l, a_{l+1}, a_{l+2}, \ldots, a_{n-1}\}$ of H. Then

$$L = \overline{\{b_1, b_2, \ldots, b_l, a_{l+1}, a_{l+2}, \ldots, a_{n-2}\}}$$

is a coline of M contained in H. By choosing $b' \in B - L, b'' \in B - H'$, we construct distinct hyperplanes $H' = \overline{L \cup b'}$ and $H'' = \overline{L \cup b''}$. Furthermore, $|H' \cap B| = |H'' \cap B| = l + 1$, hence H' and H'' are distinct from H, and $f_{H'}$ and $f_{H''}$ are in span A. But by hypothesis (2) of the proposition, since H, H' and H'' are distinct hyperplanes containing L, $f_H \in \text{span } \{f_{H'}, f_{H''}\} \subseteq \text{span } A$, completing the proof of the lemma. $\qquad\square$

Proof of Proposition 1.5.5. For any $s \in S$, we define a linear functional L_s on V by $L_s(f) = f(s) \in K$ for all $f \in V$. Then the mapping $\sigma : S \to V^*, s \to L_s$ will coordinatize $M(S)$ if we can show that independent and dependent sets are preserved under σ (since V^*, the dual space of V, is a vector space over K). Clearly it suffices to consider maximal independent sets, or bases of M, and minimal dependent sets.

Let $\{b_1, b_2, \ldots, b_n\} = B$ be any basis of $M(S)$. Then from the lemma we obtain the basis $\{f_{H_1}, f_{H_2}, \ldots, f_{H_n}\}$ of V, where $f_{H_i}(b_j) \neq 0$ if and only if $i = j$. Thus $L_{b_j}(f_{H_i}) \neq 0$ if and only if $i = j$, so $L_{b_1}, L_{b_2}, \ldots, L_{b_n}$ are independent in V^*.

Now let $\{b_0, b_1, \ldots, b_k\}$ be a minimal dependent set in $M, k \leqslant n$. Then the independent set $\{b_1, b_2, \ldots, b_k\}$ may be extended to a basis $\{b_1, b_2, \ldots, b_n\} = B$ of $M(S)$. As before, the lemma provides a basis $\{f_{H_1}, f_{H_2}, \ldots, f_{H_n}\}$ of V with $L_{b_j}(f_{H_i}) \neq 0$ if and only if $i = j$. But $b_0 \in \overline{\{b_1, b_2, \ldots, b_k\}} \subseteq \overline{B - \{b_i\}}$ for all $i > k$. Thus $L_{b_0}(f_{H_i}) = 0$ for all $i > k$. Since the linear functional L_{b_0} is determined by its values on the basis $\{f_{H_1}, f_{H_2}, \ldots, f_{H_n}\}$ of V, we have $L_{b_0} = \sum_{i=1}^{k} \alpha_i L_{b_i}$, where $\alpha_i = L_{b_0}(f_{H_i})/L_{b_i}(f_{H_i})$. Thus $L_{b_0}, L_{b_1}, \ldots, L_{b_k}$ are linearly dependent in V^*, completing the proof of the sufficiency of (1) and (2).

The converse is easy to prove. If $\zeta : S \to V$ is a coordinatization of M over K, then for any hyperplane H, $\zeta(H)$ spans a subspace U which is a hyperplane of V (that is, a subspace of dimension one less than V). Now, there is a unique (up to non-zero scalar multiple) linear functional $f_U: V \to K$ whose kernel is U, and $f_H = f_U \circ \zeta$ is the desired function, since conditions (1) and (2) may easily be checked.   □

Another sufficient condition for coordinatization, due to Kantor (1975), is that each coline has at least three hyperplanes and each rank 4 minor is coordinatizable over a fixed prime field $GF(p)$.

1.6. Brackets

Among the most useful general conditions for coordinatizability are the bracket conditions. If $\zeta: M(S) \to V$ is a coordinatization into a vector space V of dimension n over a field K, where $n = \operatorname{rank} M$, and if vectors in V are expressed as column vectors with respect to a standard basis B, then for any $x_1, x_2, \ldots, x_n \in S$, we define $[x_1, x_2, \ldots, x_n] = \det(\zeta x_1, \zeta x_2, \ldots, \zeta x_n)$. These determinants are called the brackets of ζ, and are often denoted $[X]$, where X is the sequence $(x_1, x_2, \ldots, x_n)$.

The following proposition is closely related to a result widely known to invariant theorists in the nineteenth century. This result says that assigning values to the brackets so that certain relations (called syzygies) are satisfied determines (uniquely, up to linear equivalence) a set of vectors having the assigned bracket values. Thus a map of S into V is determined simply by specifying the values of the brackets arbitrarily, provided the syzygies are satisfied. However, this classical result did not predetermine which bracket values were to be zero.

1.6.1 Proposition. *Let $M(S)$ be a matroid of rank n, and let $[x_1, x_2, \ldots, x_n]$ be assigned a value in the field K, for every $x_1, x_2, \ldots, x_n \in S$. A necessary and sufficient condition for the existence of a coordinatization ζ of M over K whose brackets are precisely the assigned values is that the following relations (or syzygies) be satisfied:*
  (1) $[x_1, x_2, \ldots, x_n] = 0$ *if and only if $\{x_1, x_2, \ldots, x_n\}$ is either dependent in M or contains fewer than n distinct elements.*
  (2) *(Antisymmetry)* $[x_1, x_2, \ldots, x_n] - (sgn\,\sigma)[x_{\sigma 1}, x_{\sigma 2}, \ldots, x_{\sigma n}] = 0$ *for every permutation σ of $\{1, 2, \ldots, n\}$, for every $x_1, x_2, \ldots, x_n \in S$.*
  (3) $[x_1, x_2, \ldots, x_n] \, [y_1, y_2, \ldots, y_n] - \sum_{i=1}^{n}[y_i, x_2, \ldots, x_n] \, [y_1, y_2, \ldots, y_{i-1}, x_1, y_{i+1}, \ldots, y_n] = 0$ *for every $x_1, \ldots, x_n, y_1, \ldots, y_n \in S$.*

Proof: We first check the necessity. Let ζ be a coordinatization of $M(S)$. From elementary properties of determinants, we see immediately that (1) and (2) are satisfied by the brackets of ζ. To verify (3), we first note that the equation is trivial unless some summand is non-zero, and hence either $\{x_1, x_2, \ldots, x_n\}$ and $\{y_1, y_2, \ldots, y_n\}$ are both bases of M, or else for some i, $\{y_i, x_2, \ldots, x_n\}$ and $\{y_1, y_2, \ldots, y_{i-1}, x_1, y_{i+1}, \ldots, y_n\}$ are both bases. In fact, we may assume the former of these, for if $\{y_i, x_2, \ldots, x_n\}$ and $\{y_1, y_2, \ldots, y_{i-1}, x_1, y_{i+1}, \ldots, y_n\}$ are both bases, the syzygy of type (3) with $[y_i, x_2, \ldots, x_n]$ $[y_1, y_2, \ldots, y_{i-1}, x_1, y_{i+1}, \ldots, y_n]$ as first term is easily checked to be equivalent to the original syzygy with $[x_1, \ldots, x_n][y_1, \ldots, y_n]$ as first term, using antisymmetry. We now apply the non-singular linear transformation $T: V \to V$ which maps ζx_j to the j-th unit vector $\mathbf{e}_j$ of V, for each j. Let $T(\zeta y_j) = \mathbf{w}_j \in V$, and let W be the $n \times n$ matrix whose j-th column is $\mathbf{w}_j$. Applying T multiplies every determinant in (3) by the same constant, hence (3) is equivalent to

$$(\det I)(\det W)$$

$$= \sum_{i=1}^{n} \det(\mathbf{w}_i, \mathbf{e}_2, \ldots, \mathbf{e}_n) \det(\mathbf{w}_1, \mathbf{w}_2, \ldots, \mathbf{w}_{i-1}, \mathbf{e}_1, \mathbf{w}_{i+1}, \ldots, \mathbf{w}_n)$$

$$= \sum_{i=1}^{n} w_{1i}(-1)^{i-1} \det W_{1i}, \tag{1.1}$$

where W_{1i} is the minor of W with row 1 and column i deleted. But equation (1.1) is just the Laplace expansion of $\det W$ by its first row. Since T is invertible, the syzygy (3) is verified.

We now prove the sufficiency. We assume that $[x_1, x_2, \ldots, x_n]$ is given as an element of K for every $x_1, x_2, \ldots, x_n \in S$ so that the syzygies are satisfied. We must construct a coordinatization ζ whose brackets are equal to the assigned values, that is

$$\det(\zeta x_1, \zeta x_2, \ldots, \zeta x_n) = [x_1, x_2, \ldots, x_n]. \tag{1.2}$$

Let $Y = \{y_1, y_2, \ldots, y_n\}$ be a basis of $M(S)$. Then $[Y] \neq 0$, and we may normalize the bracket values by dividing each of them by $[Y]$. Since the syzygies are each homogeneous, they are still satisfied by the normalized bracket values, and thus we may assume $[Y] = 1$. We now define the i-th coordinate of the vector $\zeta(x)$ by $\zeta(x)^i = [y_1, y_2, \ldots, y_{i-1}, x, y_{i+1}, \ldots, y_n]$. We will now show that $\zeta: S \to K^n$ is the desired coordinatization. Actually, it suffices to show that (1.2) holds for all $x_1, x_2, \ldots, x_n$, for then the fact that ζ is a coordinatization follows from syzygy (1).

Let $x_1, x_2, \ldots, x_n \in S$ be arbitrary. We may assume that these n elements are distinct, for otherwise $[x_1, x_2, \ldots, x_n] = \det(\zeta x_1, \zeta x_2, \ldots, \zeta x_n) = 0$. Let $X = \{x_1, x_2, \ldots, x_n\}$ and $k = |X - Y|$. We now show (1.2) by induction on k. If $k = 0$ or 1, then (1.2) holds by the definition of ζ, so suppose $k \geq 2$. Then, using the induction hypothesis,

$$[X][Y] = \sum_{i=1}^{n} [y_1, x_2, \ldots, x_n][y_1, \ldots, y_{i-1}, x_1, y_{i+1}, \ldots, y_n]$$

$$= \sum_{i=1}^{n} \det(\zeta y_i, \zeta x_2, \ldots, \zeta x_n) \det(\zeta y_1, \ldots, \zeta y_{i-1}, \zeta x_1, \zeta y_{i+1}, \ldots, \zeta y_n)$$

$$= \det(\zeta X) \det(\zeta Y)$$

since we have already verified that determinants satisfy the syzygy (3). But $[Y] = \det(\zeta Y) = 1$, hence we have proved (1.2). We may reverse the normalization by multiplying the first coordinate of every vector ζx by the original $[Y]$, thus completing the proof. ☐

An immediate corollary is one of the characterizations of binary matroids (see Chapter 2).

1.6.2. Corollary. *A matroid $M(S)$ is binary if and only if for every pair of bases X and Y of M and $x_1 \in X$, there are exactly an odd number of $y_i \in Y$ such that $Y - y_i + x_1$ and $X - x_1 + y_i$ are both bases.*

Proof. If $K = GF(2)$, the field of two elements, then according to syzygy (1), we must assign $[X] = 1$ if X is a basis, and $[X] = 0$ otherwise. Syzygy (2) is then always satisfied, and syzygy (3) is satisfied if and only if $M(S)$ satisfies the stated exchange condition. ☐

The above proposition is the foundation of the theory of the bracket ring, a tool which has proved useful in the study of several coordinatization questions, especially those relating to transcendence degree of coordinatizations, unimodular coordinatizations, and coordinatizations of rank-preserving weak-map images. The bracket ring is constructed from a polynomial ring, with an indeterminate for each bracket, by dividing by the ideal generated by all polynomials corresponding to the syzygies. Thus the 'brackets' in the bracket ring are forced to satisfy the syzygies, and Proposition 1.6.1 now says that a coordinatization of $M(S)$ over K is equivalent to a ring homomorphism of the bracket ring into K having no non-zero bracket in its kernel. This in turn is equivalent to the existence of a prime ideal in the bracket ring containing no bracket of a basis of $M(S)$. Thus many coordinatization problems may be transformed into ring-theoretic questions involving the prime ideal structure of the bracket ring (see White 1980).

An idea that is similar in spirit to the bracket ring was developed independently by Vámos (1971). He starts with an $n \times |S|$ matrix of indeterminants which he wishes to turn into a coordinatization matrix for $M(S)$. Then, in the appropriate polynomial ring, he considers the ideal I generated by all $n \times N$ determinants which correspond to non-bases of M, and the multiplicatively closed subset T generated by all $n \times n$ determinants which correspond to bases. Then $M(S)$ is coordinatizable if and only if $T \cap I = \phi$. Indeed, similarly to the bracket ring, a coordinatization corresponds to a homomorphism of the polynomial ring whose kernel contains I and does not intersect T, and hence to prime ideals which contain I and do not intersect T.

The Vámos ring has recently been further developed and its algebraic relation to the bracket ring made explicit by Fenton (1981). In particular, he obtains a ring which is a universal coordinatization ring in a stronger sense than either the bracket ring or the Vámos ring.

1.7. Coordinatization over Algebraic Extensions

The object of this section is to prove the following proposition of MacLane, which appeared in one of the earliest papers (MacLane, 1936) in matroid theory. MacLane only considered the case of characteristic zero, although the extension to arbitrary characteristic is straightforward.

Let K be a prime field, that is, K is the field of rationals of $GF(p)$ for some prime p. The proposition says, roughly, that any field L which is algebraic of finite degree over K is the unique 'minimal' coordinatizing field of some matroid.

1.7.1. Proposition. *Let L be a finite algebraic extension field of K. Then there exists a matroid M of rank 3 which may be coordinatized over L, such that if L' is any extension field of K which permits a coordinatization of M, then L' contains a subfield isomorphic to L.*

We first prove the following lemma.

1.7.2. Lemma. *Let $N(S)$ be a matroid of rank 3 with a given coordinatization ζ over a field L, which includes among its image vectors $(1, 0, 0)^t, (0, 1, 0)^t, (0, 0, 1)^t, (1, 1, 1)^t, (1, 0, a)^t$ and $(1, 0, b)^t$ for any $a, b \in L$. Then we may extend N to a matroid coordinatizable by an extension of ζ such that the image vectors in such an extension must include $(1, 0, a + b)^t, (1, 0, ab)^t, (1, 0, -a)^t$, or $(1, 0, a^{-1})^t$ if $a \neq 0$, (in each case up to scalar multiple), whichever we prefer.*

Proof.

$$
\begin{array}{ccccccccccccc}
c & d & e & f & y & z & g & h & i & s & t & u & v & w \\
\begin{pmatrix}
1 & 0 & 0 & 1 & 1 & 1 & 1 & 1 & 0 & 1 & 0 & 1 & 1 & 1 \\
0 & 1 & 0 & 1 & 0 & 0 & 1 & 0 & 1 & 1 & 1 & 0 & a & 0 \\
0 & 0 & 1 & 1 & a & b & 0 & 1 & -1 & a-b & a+b & 0 & ab
\end{pmatrix}
\end{array}
$$

Let us denote the elements of S with the given image vectors c, d, e, f, y, and z, respectively. Let us extend N by successively adjoining $g \in cd \cap ef, h \in df \cap ce, i \in gh \cap de, s \in dy \cap ef, t \in zg \cap de, u \in st \cap ce, v \in yi \cap cd, w \in tv \cap ce$. Then each of these elements must be assigned the coordinates shown up to scalar multiple in any extension of ζ. Adjoin an additional element r on ce, and construct the element u as above with r in place of z, and then identify u with c. This forces the coordinates $(1, 0, -a)^t$ to be assigned to r. A similar construction yields the multiplicative inverse $(1, 0, a^{-1})^t$ if $a \neq 0$. If not all of the four constructed vectors are desired, suitable restrictions of the constructed matroid may be employed.   □

1.7.3. Comment. The constructions we have just given are a 'geometric' analog of the algebraic operations of addition and multiplication, and their inverses, and are of considerable importance in the construction of coordinatizations of Desarguesian projective planes. The affine diagrams of these constructions, Figures 1.10 and 1.11, are illuminating. We let cd be the line at infinity, and for any vector with first coordinate equal to 1, we may take its last two coordinates as affine coordinates. We have reversed the order of these two coordinates in the following diagrams, to correspond to ordinary Cartesian coordinates. In Figure 1.10, gz and su are parallel lines (meeting at the point t on the line at infinity). Their slope is $-1/b$. In Figure 1.11, gh and vy are parallel, of slope -1, and gz and vw are parallel of slope $-1/b$.

Figure 1.10. Geometric addition.

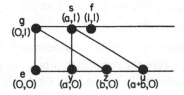

Figure 1.11. Geometric multiplication.

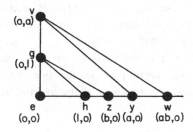

Proof of Proposition 1.7.1. From field theory, since L is either finite or of characteristic zero, we know that there exists a primitive element α, that is, $\alpha \in L$ such that $L = K[\alpha]$. Let $p(x)$ be the minimal polynomial of α over K. We will now construct a matroid M which is coordinatizable over L, such that a root of $p(x)$ is required in any extension field L' of K over which it may be coordinatized. The proposition will then follow from standard field theory.

Beginning with the five vectors $(1,0,0)^t, (0,1,0)^t, (0,0,1)^t, (1,1,1)^t, (1,0,1)^t$ in a three-dimensional vector space V over L, we can construct a vector $(1,0,a)^t$ for any integer a in K, by repeatedly adding 1 to itself and by using additive inverse. Since K is a prime field, we may assume that $p(x)$ has integral coefficients.

Now we adjoin $(1, 0, \alpha)^t$, and by successive use of the addition and multiplication constructions, adjoin all elements necessary to construct $p(\alpha)$. But since we are working in a vector space over L, $p(\alpha) = 0$, thus the last vector in our construction, $(1, 0, p(\alpha))^t$, is in fact $(1, 0, 0)^t$. We now check that the matroid M consisting of all the vectors used in the construction is the required matroid. It has rank 3 and is *a priori* coordinatizable over L. Let L' be an extension field of K over which M may be coordinatized. By putting this coordinatization into projective canonical form, we may assume that $(1, 0, 0)^t, (0, 1, 0)^t, (0, 0, 1)^t$, and $(1, 1, 1)^t$ (of the original coordinatization over L) are assigned the same coordinates over L'. But then our addition construction forces $(1, 0, a)^t$ to be assigned the same coordinates over L' for any integer $a \in K$. Finally $(1, 0, \alpha)^t$, since it is collinear with $(1, 0, 0)^t$ and $(0, 0, 1)^t$, must be assigned coordinates $(1, 0, \beta)^t$, for some $\beta \in L'$. But we have all the vectors necessary for the construction of $(1, 0, p(\beta))^t$, and we know that this element of our matroid is equal to $(1, 0, 0)^t$. Hence β is a root of $p(x)$, and since $p(x)$ is irreducible, standard field theory tells us there is an isomorphism of L into L', by taking the identity map on K and mapping α to β. □

1.7.4. Remark. The construction of M was accomplished inside a vector space over L in order to assure the coordinatizability of M over L. Although the arithmetic constructions can be carried out on an abstract matroid, there may be additional dependencies not accounted for in the construction which are necessary for coordinatization over L. As an example of this, if $p(x) = x^2 + x + 1$, then we would use the vectors $(1, \alpha, 0)$ and $(0, 1, -\alpha^2)^t$ in our construction, but $\alpha^2 + \alpha + 1 = 0$ forces these two vectors to be collinear with $(1, 0, 1)^t$, since $\alpha^3 - 1 = (\alpha - 1)(\alpha^2 + \alpha + 1) = 0$.

These constructions may also be used to provide examples of matroids coordinatizable only over certain characteristics. Example 1.2.3 may be regarded as the geometric construction of the arithmetic statement '$2 \neq 0$'.

1.8. Characteristic Sets

Let $P = \{p : p \text{ is a prime number}\} \cup \{0\}$. The *characteristic set* $C(M)$ of a finite matroid M is the subset of P consisting of the characteristics of fields which coordinatize M. Ingleton (1971) raised the problem of determining which subsets of P are characteristic sets. The existence of unimodular and non-coordinatizable matroids shows that P and ϕ are characteristic sets. Rado (1957) proved that $0 \in C(M) \Rightarrow C(M)$ is cofinite, i.e., $C(M)$ includes all but a finite number of primes. Vámos (1971) showed that $0 \notin C(M) \Rightarrow C(M)$ is finite. Reid (unpublished) showed that all cofinite subsets of P which include 0 are characteristic sets, leaving open only the question of finite characteristic sets. Recently Jeff Kahn (1981) has completely settled the problem by proving that all finite subsets of P which do not include 0 are characteristic sets.

1.8.1. Proposition. *If $Q \subseteq P$, then $Q = C(M)$ for some finite matroid M if and only if either Q is cofinite and includes 0, or Q is finite and excludes 0.*

Kahn's proof for finite subsets Q is rather interesting. If $Q = \{p_1, p_2, \ldots, p_k\}$, let $N = p_1 p_2 \ldots p_k$. We wish to construct a matroid using the methods of the previous section (but on abstract matroids rather than within a particular vector space), in such a way that the vector $(1, 0, N)^t$ must be used; we then identify that point of the matroid with the point assigned the vector $(1, 0, 0)^t$, thus forcing the algebraic relation $N = 0$. Kahn actually works with a multiple of N which is of the form $2^k - 1$. This idea is certainly not new; however there are several difficulties to overcome. For example, as intermediate powers 2^j are formed on the way to 2^k, some of the primes in Q may divide $2^j - 1$. Then $(1, 0, 2^j)^t = (1, 0, 1)^t$ in a coordinatization over such primes, but we want these to be distinct points. Another difficulty is that 'random' collinearities such as mentioned in Remark 1.7.4 may occur over one prime $p_i \in Q$ but not over another $p_j \in Q$. Kahn has devised ingenious remedies to these difficulties. For example, he may require three lines, l_1, l_2, l_3, to be coincident at some point x, but the point x is collinear with another line l over p_i but not over p_j. He then replaces the point x by a Desargues configuration, as shown in Figure 1.12, which forces the coincidence of l_1, l_2, l_3 in any coordinatization, but since x is no longer a point in the matroid, its collinearity with l is no longer an issue.

Figure 1.12. Kahn's trick.

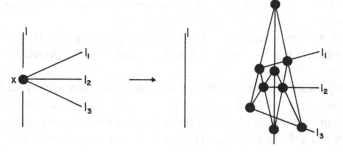

The actual construction of a reasonably small matroid with a given finite characteristic set is far from easy, if at least two primes are involved. Lazarson (1958) provided the construction for singletons $\{p\}$. Reid (unpublished) constructed a matroid for $\{1103, 2089\}$ (see Exercise 1.6).

1.9. Coordinatizations over Transcendental Extensions

We might logically ask, after having considered coordinatizations over algebraic extensions of fields, when a matroid has a coordinatization over an arbitrary extension of a given field. Our first result, due to MacLane (1936) is that it is never necessary to use a non-algebraic extension.

1.9.1. Proposition. *Let K be a field and $M(S)$ a finite matroid which is coordinatizable over an extension field L over K. Then there exists an algebraic extension field L' over K such that M is coordinatizable over L'.*

Nevertheless, coordinatizations using transcendentals are very useful. Transversal matroids may be characterized using transcendental coordinatizations (see Theorem 5.4.7). We have seen in Chapter 7 of White (1986) that coordinatizations for many of the matroid constructions may be conveniently obtained using transcendentals, including principal extension, principal lift, truncation, Dilworth truncation, and matroid union.

It is often useful to coordinatize a matroid using transcendentals, and then later specialize the transcendentals. Let $M(S)$ be coordinatized over $L = K(x_1, x_2, \ldots, x_n)$ where $x_1, x_2, \ldots, x_n$ are transcendental over K, and suppose that every entry in the coordinatization matrix is in the subring $R = K[x_1, x_2, \ldots, x_n]$ of L. That is, $\zeta : S \to L^n, \zeta(S) \subseteq R^n \subseteq L^n$. Then if L' is any field extension of K, we can define a ring homomorphism $f : R \to L'$ by specifying the images of $x_1, x_2, \ldots, x_n$, and mapping K identically to K. If $f^n : R^n \to (L')^n$ is the map induced by applying f coordinatewise, and if the composition $f^n \circ \zeta$ is a coordinatization of M, we call this coordinatization a specialization of ζ. Analogous definitions may be made if we take R to be $\mathbb{Z}[x_1, x_2, \ldots, x_n]$ where $\mathbb{Z}$ is the ring of integers; then a coordinatization may be specialized to coordinatizations of distinct characteristics.

A number of matroids have the interesting property that they possess a coordinatization which is universal with respect to such specializations.

1.9.2. Proposition. *Let M be a unimodular matroid and B a basis of M. Then M has a coordinatization ζ_0 over $\mathbb{Z}[x_1, x_2, \ldots, x_n]$, for appropriate m, which is in echelon form with respect to B, such that every coordinatization ζ of M over any field, with ζ in echelon form with respect to B, is a specialization of ζ_0.*

Proof. Let $(I \mid A)$ be any unimodular coordinatization of M over $\mathbb{Z}$, in echelon form with respect to B. We obtain ζ_0 by multiplying the rows and columns of A by distinct indeterminants $x_1, x_2, \ldots, x_m$. Now if ζ is any coordinatization of M over a field K, in echelon form with respect to B, it is projectively equivalent to (I, A) by Proposition 1.2.5, and hence may be obtained from (I, A) (with entries viewed as in K) by scalar multiplications on the rows and columns of A. Hence ζ is a specialization of ζ_0. $\qquad\qquad\square$

It is easy to check that principal transversal matroids (Brylawski 1975, White 1986, Proposition 7.4.2, part 3) also have such universal coordinatizations, and so do many other matroids. It is not known, however, whether the class of transversal matroids always have such universal coordinatizations.

1.9.3. Example. The 9-point planar matroid illustrated in Figure 1.13 is transversal with the given coordinatization using 18 indeterminates corresponding to its maximal transversal presentation:

$$\begin{pmatrix} x_1 & x_2 & x_3 & x_7 & x_8 & x_9 & 0 & 0 & 0 \\ x_4 & x_5 & x_6 & 0 & 0 & 0 & x_{13} & x_{14} & x_{15} \\ 0 & 0 & 0 & x_{10} & x_{11} & x_{12} & x_{16} & x_{17} & x_{18} \end{pmatrix}$$

Figure 1.13. A 9-point transversal matroid.

However, this coordinatization cannot be specialized to any coordinatization in which the three non-trivial lines are coincident in the ambient space. But this matroid has another coordinatization where y_1, y_2, y_3 are chosen to make the three 3×3 minors dependent. Although y_1, y_2, y_3 may be written as rational functions in the x_1's, denominators may be cleared to obtain a representation in $\mathbb{Z}[x_1, x_2, \ldots, x_{24}]$ which is universal with respect to specializations:

$$\begin{pmatrix} x_1 & x_2 & x_3 & x_7 & x_8 & x_9 & x_{23} & x_{24} & y_3 \\ x_4 & x_5 & x_6 & x_{21} & x_{22} & y_2 & x_{13} & x_{14} & x_{15} \\ x_{19} & x_{20} & y_1 & x_{10} & x_{11} & x_{12} & x_{16} & x_{17} & x_{18} \end{pmatrix}$$

1.10. Algebraic Representation

An interesting alternative to coordinatization (linear representation) of a matroid is algebraic representation. Let K be a field of finite transcendence degree over the field k. It is well-known (MacLane, 1938) that K forms a matroid by defining independence to mean algebraic independence over k. This provides another interesting special class of matroids, and raises the natural problem of determining when a given matroid M is isomorphic to such an algebraic independence matroid. We assume that M is finite.

Thus we say that $\zeta : S \to K$ is an *algebraic representation of $M(S)$ over k* if $A \subseteq S$ is independent in M if and only if $\zeta|_A$ is injective and $\zeta(A)$ is algebraically independent over k. Equivalently, by restricting our attention to the subfield K' of K generated by $\zeta(S)$, an algebraic representation ζ of M may also be characterized by: $B \subseteq S$ is a basis of M if and only if $\zeta|_B$ is injective and $\zeta(B)$ is a transcendence basis of K'/k. If there exists such an algebraic representation of M over k, we say that M is *algebraic over* k.

1.10.1. Proposition. *If* M *is coordinatizable over* k, *then* M *is algebraic over* k.

Proof. Let $\zeta: S \to V$ be a coordinatization of M over k. Pick a basis $\{b_i\}_{i=1}^n$ of V and an algebraically independent set $\{t_i\}_{i=1}^n$ in an appropriate extension K of k and define $\phi: V \to K$ by $\phi(\sum \alpha_i b_i) = \sum \alpha_i t_i$. Then the composition $\phi \circ \zeta$ is the desired algebraic representation of M, as the reader can easily verify. $\square$

The converse is not true, as shown by Counterexample 1.10.6, but in characteristic 0 the converse is true.

1.10.2. Proposition. (*See, for example, Lang 1965, Chapter 10, Proposition 10*) *If* M *is algebraic over a field* k *of characteristic* 0 *then* M *is coordinatizable over some finite transcendental extension of* k.

1.10.3. Proposition. (*Lindström 1985a*) *If* M *is algebraic over the transcendental extension* $k(T)$ *of* k, *then* M *is algebraic over* k.

1.10.4. Corollary. *If* M *is algebraic over a field* k *of characteristic* 0 *then* M *is coordinatizable over* $\bar{k}$, *where* $\bar{k}$ *is the algebraic closure of* k.

1.10.5. Corollary. (*Lindström 1985a*) *If* M *is algebraic over* k *then* M *is algebraic over the prime field of* k.

1.10.6. Counterexample. (*Lindström 1985b*) *The non-Pappus matroid* (*see Figure 1.7*) *is algebraic over any finite field.*

As we have previously noted, the non-Pappus matroid (Figure 1.7) is non-coordinatizable over every field, and is also non-algebraic over every field of characteristic 0, by Corollary 1.10.4.

Several examples of matroids non-algebraic over every field are known, including the non-Desargues matroid (Lindström 1984) and the Vámos cube (Ingleton and Main 1975). See Figures 1.6 and 1.8.

In analogy with characteristic sets $C(M)$ for linear representation of matroids, as discussed in Section 1.8, we may consider characteristic sets for algebraic representation. Thus for a matroid M, we define the *algebraic characteristic set*, $A(M)$, to be the subset of $P = \{\text{primes}\} \cup \{0\}$ consisting of the characteristics of fields over which M has an algebraic representation.

The results above show that: (1) for all matroids, $C(M) \subseteq A(M)$; (2) $0 \in A(M) \Rightarrow 0 \in C(M)$; (3) $0 \in A(M) \Rightarrow A(M)$ is cofinite. The only known cofinite algebraic characteristic sets are P and $P - \{0\}$ (e.g., non-Pappus). All singletons $\{p\}$ except $\{0\}$ are algebraic characteristic sets, specifically for the

Lazerson matroids, which, as we have noted, also have $\{P\}$ for their linear characteristic sets (Lindström 1985c). A number of finite, non-singleton, algebraic characteristic sets are known (Gordon 1987).

It is not hard to show the following proposition (see Welsh 1976, p. 187).

1.10.7. Proposition. *If $M(S)$ is algebraic over F and $A \subseteq S$, then the contraction M/A is algebraic over a transcendental extension of F, and hence over F.*

1.10.8. Corollary. *If M is algebraic over F, then so is every minor of M.*

An obvious question then is to investigate excluded minor characterizations for algebraic representation over various fields. Very little has been done on this problem.

It is known (Welsh 1976) that the class of algebraic matroids is closed under truncations and matroid unions. The obviously important question of whether it is closed under duality is still open.

We close this section with some examples. The set $\{x, y, z, x + y, x + z, y + z, x + y + z\}$ in $F(x, y, z)$ algebraically represents the Fano matroid (see Figure 1.9) if F is of characteristic 2, and the non-Fano matroid (see Figure 1.14) otherwise, where x, y, and z are algebraically independent transcendentals over F. In both cases, the algebraic representation is just the image of a linear representation *via* Proposition 1.10.1. On the other hand, $\{x, y, z, xy, xz, yz, xyz\}$ represents the non-Fano matroid over all fields F. Ingleton (1971) combined these two over F of characteristic 2 to provide a matroid with 11 elements which is algebraic over F of characteristic 2 but is linear over no field.

Figure 1.14. The non-Fano matroid.

Finally we provide an algebraic representation of the non-Pappus matroid over $GF(2)$, due to Lindström (1983). As we have noted, this matroid is actually algebraic over any finite field. With the points denoted as in Figure 1.7, we let

$$\zeta(1) = x, \zeta(2) = x + y, \zeta(3) = y, \zeta(4) = \frac{xz}{x + y} + x + y, \zeta(5) = z,$$

$$\zeta(6) = \frac{yz}{x + y} + x + y, \zeta(\mathbf{x}) = xz, \zeta(\mathbf{y}) = \frac{xyz}{x + y} + xy, \zeta(\mathbf{z}) = yz.$$

It must be arduously checked that every triple of collinear points is mapped by

ζ to algebraically dependent elements of $GF(2)(x, y, z)$, and that every non-collinear triple is mapped to algebraically independent elements.

Exercises

1.1. If M is coordinatizable, characterize the hyperplanes of M in terms of the placement of zeros in coordinatizations of M.

1.2. Characterize the circuits of a vector matroid, and show directly that they satisfy the various circuit elimination axioms.

1.3. Construct a matroid M requiring the golden mean, $\alpha = (1 + \sqrt{5})/2$, to be an element of F if F is a subfield of $\mathbb{R}$ over which M may be coordinatized.

1.4. Verify that the *symmetric subset basis exchange axiom* holds for coordinatizable matroids. This axiom states: for all B, B', A such that $B \in \mathcal{B}$, $B' \in \mathcal{B}$, $A \subseteq B$, there exists $A' \subseteq B'$ such that $(B - A) \cup A' \in \mathcal{B}$, $(B' - A') \cup A \in \mathcal{B}$.

1.5. Prove algebraically, by directly trying to construct a coordinatization, that the non-Desargues configuration and the Vámos matroid are both non-coordinatizable.

1.6. Construct a matroid which may be coordinatized over the field K if and only if the characteristic of K is 1103 or 2089. (*Hint*: $2^{29} - 1 = 1103 \cdot 2089 \cdot 233$.)

1.7. Let R_{10} be the binary matroid with binary coordinatization as shown. Show that R_{10} is self dual. Show that all minors of R_{10} are either graphic or cographic.

$$\begin{pmatrix} 1 & 0 & 0 & 0 & 0 & 1 & 1 & 1 & 0 & 0 \\ 0 & 1 & 0 & 0 & 0 & 0 & 1 & 1 & 1 & 0 \\ 0 & 0 & 1 & 0 & 0 & 0 & 0 & 1 & 1 & 1 \\ 0 & 0 & 0 & 1 & 0 & 1 & 0 & 0 & 1 & 1 \\ 0 & 0 & 0 & 0 & 1 & 1 & 1 & 0 & 0 & 1 \end{pmatrix}$$

1.8. Prove that principal transversal matroids have coordinatizations which are universal with respect to specialization, for coordinatizations in echelon form with respect to the canonical basis.

1.9. Prove that if F_7 or F_7^* is coordinatizable over the field K, then K must have characteristic 2.

1.10. Prove that F_7^* has, up to isomorphism, only two distinct binary rank-4 1-element extensions, one of which is $AG(3, 2)$. Each of these two matroids is isomorphic to its own orthogonal matroid.

1.11. Construct a matroid which may be coordinatized over characteristic p, for a given prime number p, but over no other characteristic.

1.12. Let M be the 10-element matroid coordinatized by the given matrix over $\mathbb{Q}$. Determine an algebraic representation of M over $GF(2)$.

$$\begin{pmatrix} 1 & 0 & 0 & 1 & 1 & 0 & 1 & 1 & 0 & 1 \\ 0 & 1 & 0 & 1 & 0 & 1 & -1 & 0 & 1 & 1 \\ 0 & 0 & 1 & 0 & 1 & 1 & 0 & -1 & -1 & 1 \end{pmatrix}$$

1.13. (Ingleton and Main 1975) Show that the Vámos matroid (Figure 1.8) is not algebraic.

References

Bixby, R.E. (1979). On Reid's characterization of the matroids representable over $GF(3)$. *J. Comb. Theory Ser. B* **26**, 174–204.

Brylawski, T. (1975). An affine representation for transversal geometries. *Studies in Applied Math.* **54**, 143–60.

Brylawski, T. and Lucas, D. (1973). Uniquely representable combinatorial geometries, in *Proceedings of the International Colloquium on Combinatorial Theory*, Rome.

Greene, C. (1971). Lectures on combinatorial geometries, Advanced Science Seminar in Combinatorial Theory, Bowdoin College, Bowdoin, Maine.

Fenton, N. (1981). Representations of matroids. Ph.D. Thesis, University of Sheffield.

Gordon, G. (1987). Algebraic characteristic sets of matroids. *J. Comb. Theory Ser. B.* to appear.

Ingleton, A.W. (1971). Representation of matroids, in *Combinatorial Mathematics and its Applications*, pp. 149–69. Academic Press, London and New York.

Ingleton, A. and Main, R. (1975). Non-algebraic matroids exist. *London Math. Soc.* **7**, 144–6.

Kahn, J. (1981). Characteristic sets of matroids. Dennison–OSU Math. Conference, Dennison, Ohio.

Kantor, W.M. (1975). Envelopes of geometric lattices. *J. Comb. Theory Ser. A* **18**, 12–26.

Lang, S. (1965). *Algebra*, Addison-Wesley.

Lazarson, T. (1958). The representation problem for independence functions. *J. London Math. Soc.* **33**, 21–5.

Lindström, B. (1983). The non-Pappus matroid is algebraic. *Ars Combinatoria*, **16-B**, 95–6.

Lindström, B. (1984). On algebraic representations of matroids. *Department of Math., Univ. of Stockholm, Reports No. 9.*

Lindström, B. (1985a). More on algebraic representations of matroids. *Proc. Amer. Math. Soc.* to appear.

Lindström, B. (1985b). The non-Pappus matroid is algebraic over any finite field. *Utilitas Mathematica*, to appear.

Lindström, B. (1985c). A class of algebraic matroids with simple characteristic set. *Proc. Amer. Math. Soc.* **95**, 147–51.

MacLane, S. (1936). Some interpretations of abstract linear dependence in terms of projective geometry. *Amer. J. Math.* **58**, 236–40.

MacLane, S. (1938). A lattice formulation for transcendence degrees and p-bases. *Duke Math. J.* **4**, 455–68.

Rado, R. (1957). Note on independence functions. *Proc. London Math. Soc.* **7**, 300–20.

Seymour, P.D. (1979). Matroid representation over $GF(3)$. *J. Comb. Theory Ser. B* **26**, 159–73.

Seymour, P.D. (1980). Decomposition of regular matroids. *J. Comb. Theory Ser. B* **28**, 305–60.

Tutte, W.T. (1958). A homotopy theorem for matroids, I, II. *Trans. Am. Math. Soc.* **88**, 144–74.

Tutte, W.T. (1959). Matroids and Graphs, *Trans. Am. Math. Soc.* **90**, 527–52.

Vámos, P. (1971). A necessary and sufficient condition for a matroid to be linear. *Mobius Algebra Conference, University of Waterloo*, 166–73.

Welsh, D.J.A. (1976). *Matroid Theory*. Academic Press, London.

White, N. (1971). The bracket ring and combinatorial geometry. Thesis, Harvard University.

White, N. (1980). The transcendence degree of a coordinatization of a combinatorial geometry. *J. Comb. Theory Ser. B* **29**, 168–75.

White, N., ed. (1986). *Theory of Matroids*. Cambridge University Press.

White, N., ed. (1988). *Combinatorial Geometries: Advanced Theory*. Cambridge University Press, to appear.

2

Binary Matroids

J.C. FOURNIER

Binary matroids play an important theoretical role, partly because they were the first class of coordinatizable matroids to be completely characterized, but also because the class of binary matroids contains the unimodular matroids and the graphic matroids, two classes fundamental to matroid theory. There are numerous characterizations of binary matroids, very different in nature, and expressive of the richness of the concept.

2.1. Definition and Basic Properties

2.1.1. Definition. A matroid is *binary* if it is representable (coordinatizable) over the two-element field $GF(2)$.

According to the general definition of representable matroids, (see Chapter 1), a matroid $M(E)$ on a finite set E is binary if there is a mapping α of E into a $GF(2)$-vector space V such that a subset $X \subseteq E$ is independent in $M(E)$ if and only if the restriction of α to X is injective and the set $\{\alpha(x) \mid x \in X\}$ of vectors in V is linearly independent. The mapping α is then called a *binary representation* of the matroid $M(E)$.

2.1.2. Example. Denote by $U_{r,n}$ up to isomorphism, the matroid on a set of n elements, in which the bases are those subsets which have r elements. Then $U_{2,3}$ is binary. (This matroid is identified with the projective line over the field $GF(2)$.) On the other hand, $U_{2,4}$ is not binary; this matroid, which consists of four geometric points on a line, is a typical non-binary matroid, and serves to characterize the binary matroids, as we shall see later.

2.1.3. Proposition. *If a matroid M is binary, each of its minors is binary.*

2.1.4. Proposition. *If a matroid M is binary, its orthogonal matroid M^* is binary.*

These propositions are special cases of general theorems concerning matroids representable over a field. But we can prove them directly. The first is quite trivial; the second follows from the equivalence of conditions (0) and (6) in Theorem 2.2.1, below.

We will use $\triangle$ to denote symmetric difference of sets.

2.2. Characterizations of Binary Matroids

2.2.1. Theorem. *The following are equivalent conditions concerning a matroid* $M(E)$:

(0) $M(E)$ *is binary.*

(1) *For every basis B and every circuit C, if we let $C(x)$ denote the fundamental circuit formed by an element $x \in C \backslash B$ with respect to the basis B,*

$$C = \underset{x \in C \backslash B}{\triangle}\, C(x),$$

that is, C is the mod 2 sum of those circuits.

(2) *Given any k circuits $C_1, C_2, \ldots, C_k$, their symmetric difference $C_1 \triangle C_2 \triangle \cdots \triangle C_k$ is a disjoint union of circuits (perhaps empty).*

(3) *Given any k circuits $C_1, C_2, \ldots, C_k$, their symmetric difference $C_1 \triangle C_2 \triangle \cdots \triangle C_k$ is either empty, or it contains a circuit.*

(4) *Given any two distinct circuits C_1, C_2, then $C_1 \triangle C_2$ contains a circuit.*

(5) *Given any two circuits C_1, C_2, which form a modular pair, where C_1 and C_2 are distinct but not disjoint, then $C_1 \triangle C_2$ is a circuit.*

(6) *For every circuit C and every cocircuit (or bond) C^*, $|C \cap C^*|$ is even.*

(7) *$M(E)$ contains no minor isomorphic to the matroid $U_{2,4}$ (the geometry of four points on a line).*

(8) *Every coline is contained in at most three copoints (hyperplanes).*

(9) *Given any two bases B_1 and B_2, and an element $y \in B_2$, there are an odd number of elements $x \in B_1$ such that $B_1 - x + y$ and $B_2 - y + x$ are bases.*

(10) *Given any two distinct circuits C_1 and C_2, and two elements a and b of $C_1 \cap C_2$, there is a circuit $C_3 \subseteq (C_1 \cup C_2) \backslash \{a, b\}$.*

Conditions (1) through (4), which appear in various forms in Whitney (1935), Rado (1957), Tutte (1965), Lehman (1964), and Minty (1966), express characteristic properties of the cycle space of binary matroids. Condition (5), concerning modular pairs of circuits, is due to N.L. White (1971). Condition (9) expresses a property of 'syzygies' due to G.-C. Rota and C. Greene, as described in the proof of Corollary 1.6.2. Condition (10), more recent, strengthens the axiom of elimination between two circuits in the form of an axiom of 'double elimination' (due to J.-C. Fournier).

By Proposition 2.1.4, each of the conditions of the theorem, when applied to the dual matroid $M^*(E)$, is also a property characteristic of binary matroids.

Thus, by duality, condition (8) of the theorem corresponds to the Tutte (1965) condition: *every line has at most three points* ('line' and 'point' are here taken in the sense of Tutte 1965), a condition which is expressed somewhat differently by condition (5) of the theorem (we see the duality between conditions (8) and (5) again in the proof of the theorem). Condition (6), due to Minty, is of particular interest in that it is identical to its own dual condition: we say it is *self-dual*.

Conditions (7) and (8) concern excluded configurations. They can be interpreted in the language of lattices. For example, condition (7) gives the following property of the lattice of flats of the matroid: *every interval of length 2 has at most five elements.*

The following proof of the theorem makes use of some minor technical results concerning modular pairs of circuits. Since these results are not of immediate concern in the context of binary matroids, they are relegated to an appendix to this chapter.

Proof of the theorem. We will show first of all the equivalence of conditions (1), (2), (3), (4), (5), (6), (9), and (10), (those which involve circuits) by establishing the implications

$$(3) \Rightarrow (2) \Rightarrow (4) \Rightarrow (10) \Rightarrow (5) \Rightarrow (1) \Rightarrow (6) \Rightarrow (9) \Rightarrow (6) \Rightarrow (3).$$

We then show the equivalence of these conditions with (0), by proving $(3) \Rightarrow (0) \Rightarrow (4)$. Finally we show the equivalence with (7) and (8) by proving $(0) \Rightarrow (7) \Rightarrow (8) \Rightarrow (0)$.

$(3) \Rightarrow (2)$:

If the set $C_1 \triangle C_2 \triangle \cdots \triangle C_k$ is not empty, it contains a circuit D_1, and we have

$$(C_1 \triangle C_2 \triangle \cdots \triangle C_k) \backslash D_1 = C_1 \triangle C_2 \triangle \cdots \triangle C_k \triangle D_1,$$

a set which in turn, if it is not empty, contains a circuit D_2 disjoint from D_1, and we have

$$(C_1 \triangle C_2 \triangle \cdots \triangle C_k \triangle D_1) \backslash D_2 = C_1 \triangle C_2 \triangle \cdots \triangle C_k \triangle D_1 \triangle D_2.$$

The sets under consideration being finite, we have thus a finite number of mutually disjoint circuits $D_1, \ldots, D_l$, such that

$$C_1 \triangle C_2 \triangle \cdots \triangle C_k \triangle D_1 \triangle \cdots \triangle D_l = \varnothing,$$

and thus such that $C_1 \triangle C_2 \triangle \cdots \triangle C_k = D_1 + \cdots + D_l$ (disjoint union).

$(2) \Rightarrow (4)$:

This is trivial, by the observation that $C_1 \triangle C_2 \neq \varnothing$ because $C_1 \neq C_2$.

$(4) \Rightarrow (10)$:

Trivial.

$(10) \Rightarrow (5)$:

Given a modular pair of distinct circuits C_1 and C_2, and an element $a \in C_1 \cap C_2$, we know in general that there is a *unique* circuit C_3 such that $C_3 \subseteq (C_1 \cup C_2) \setminus \{a\}$, and that moreover, $C_3 \supseteq C_1 \triangle C_2$ (see appendix). Thus, in this case if $C_3 \neq C_1 \triangle C_2$, the existence of an element $b \in C_3 \setminus (C_1 \triangle C_2)$, that is, such that $b \in C_1 \cap C_2$ and $b \in C_3$, and the uniqueness of C_3 contradict condition (10). Thus $C_3 = C_1 \triangle C_2$.

$(5) \Rightarrow (1)$:

We reason by induction on the number of elements in the set $C \setminus B$. If $|C \setminus B| = 1$, the required condition is trivial. Suppose then that $|C \setminus B| \geqslant 2$. There exist two circuits C_1 and C_2 forming a modular pair, such that $C_1 \neq C_2$, $C_1 \cap C_2 \neq \varnothing$, and $C_1 \triangle C_2 \subseteq C$ (see appendix). Since, according to (5), $C_1 \triangle C_2$ is a circuit, we have in fact $C = C_1 \triangle C_2$, and consequently the sets $C_1 \setminus B$ and $C_2 \setminus B$ partition $C \setminus B$. Furthermore, since $|C_1 \setminus B| < |C \setminus B|$ and $|C_2 \setminus B| < |C \setminus B|$, we have, by the induction assumption,

$$C_1 = \mathop{\triangle}_{x \in C_1 \setminus B} C(x) \quad \text{and} \quad C_2 \mathop{\triangle}_{x \in C_2 \setminus B} C(x),$$

whence

$$C = C_1 \triangle C_2 = \mathop{\triangle}_{x \in C \setminus B} C(x).$$

$(1) \Rightarrow (6)$:

Given any cocircuit (bond) C^* of a matroid, there exists a basis B such that $|B \cap C^*| = 1$, as it is easy to see. Let $B \cap C^* = \{y\}$. For any element $x \notin B$ we have $y \in C(x)$ if and only if $x \in C^*$ (C^* is the fundamental bond containing y, relative to the cobasis $E \setminus B^*$)[†]. Consequently, $C(x) \cap C^* = \varnothing$ or $\{x, y\}$. Hence the set

$$C \cap C^* = \left(\mathop{\triangle}_{x \in C \setminus B} C(x) \right) \cap C^* = \mathop{\triangle}_{x \in C \setminus B} (C(x) \cap C^*)$$

is of even cardinality, being the symmetric difference of sets of even cardinality.

$(6) \Rightarrow (9)$:

Note that in (9), the expression $B_1 - x + y$ means $(B_1 \setminus \{x\}) \cup \{y\}$. When $y \in B_1$, the condition (9) is trivially verified, because in that case x and y must be equal. So suppose that $y \notin B_1$ and denote by $C_{B_1}(y)$ the fundamental circuit formed by y with respect to the basis B_1, and by $C_{B_2}^*(y)$ the fundamental cocircuit formed by y relative to the cobasis $E \setminus B_2$. It is easy to show that $B_1 - x + y$ is a basis if and only if $x \in C_{B_1}(y)$, and similarly that $B_2 - y + x$ is a basis if and only if $x \in C_{B_2}^*(y)$. Since the set $C_{B_1}(y) \cap C_{B_2}^*(y)$ has an even number of elements, and moreover has only the element y not in B_1 [as does $C_{B_1}(y)$], there

---

[†] In general, in a matroid $M(E)$, given a basis B, we have the following principle of reciprocity between circuits and cocircuits which is, moreover, easy to prove: for any elements $x \in E \setminus B$ and $y \in B$, $y \in C(x) \Leftrightarrow x \in C^*(y)$.

are thus an odd number of elements

$$x \in B_1 \cap C_{B_1}(y) \cap C^*_{B_2}(y),$$

that is, of elements x satisfying condition (9).

$(9) \Rightarrow (6)$:

Let C and C^* be any circuit and cocircuit, respectively. We can assume they have a non-empty intersection, and let $y \in C \cap C^*$. There is a basis $B_1 \supset C - y$ and a cobasis $B^*_2 \supset C^* - y$; let $B_2 = E \backslash B^*_2$. Thus we have $C = C_{B_1}(y)$, the fundamental cycle of y with respect to B_1, and $C^* = C^*_{B_2}(y)$, the fundamental cocycle of y with respect to $E \backslash B_2$. Condition (9), applied to B_1, B_2, and y, taking into account the necessary and sufficient conditions given in the preceding proof in order that $B_1 - x + y$ and $B_2 - y + x$ be bases, shows that $|B_1 \cap C_{B_1}(y) \cap C^*_{B_2}(y)|$ is odd, and, on adjoining the element y, $|C_{B_1}(y) \cap C^*_{B_2}(y)| = |C \cap C^*|$ is even.

$(6) \Rightarrow (3)$:

If the set $A = C_1 \triangle \cdots \triangle C_k$ were non-empty and independent, there would exist a cocircuit C^* such that $|C^* \cap A| = 1$, as it is easy to see (consider a fundamental cocircuit relative to a cobasis disjoint from A). We then reach a contradiction because $C^* \cap (C_1 \triangle \cdots \triangle C_k) = (C^* \cap C_1) \triangle \cdots \triangle (C^* \cap C_k)$ is a set of even cardinality, the sets $C^* \cap C_i$ all being even.

$(3) \Rightarrow (0)$:

In the $GF(2)$ vector space $\mathscr{P}(E)$ of subsets of the set E (addition in this space being symmetric difference, and scalar product being trivial), let Γ be the subspace spanned by the circuits of the matroid $M(E)$. For any element $x \in E$, let $\alpha(x)$ be the congruence class of x modulo Γ in $\mathscr{P}(E)$. There is no difficulty in showing that the map α of E into the $GF(2)$ quotient space $V = \mathscr{P}(E)/\Gamma$ is a binary linear representation of $M(E)$ (see the definition).

$(0) \Rightarrow (4)$:

Let α be a binary representation of $M(E)$ in a vector space V over $GF(2)$. Given any circuit C in $M(E)$, we show we have the 'dependence relation' $\sum_{x \in C} \alpha(x) = 0$. Given any two circuits C_1 and C_2, by forming the sum of the corresponding dependence relations we find $\sum_{x \in C_1 \triangle C_2} \alpha(x) = 0$, a relation which shows that $\alpha(C_1 \triangle C_2)$ is dependent in V and thus that $C_1 \triangle C_2$ is dependent in $M(E)$ and must contain a circuit.

$(0) \Rightarrow (7)$:

This is an immediate consequence of Proposition 2.1.3, and of the fact that the matroid $U_{2,4}$ is not binary. [We see, for example, that it does not satisfy condition (4).]

$(7) \Rightarrow (8)$:

If there were to exist in $M(E)$ a coline L contained in four hyperplanes (copoints) H_1, H_2, H_3, H_4, there would be a minor $M(H_1 \cup H_2 \cup H_3 \cup H_4)/L$ isomorphic to $U_{2,4}$.

$(8) \Rightarrow (0)$:

In fact we will show that $(8) \Rightarrow (5)^*$, the dual of condition (5). This is sufficient because if $M(E)$ satisfies $(5)^*$, that is, if $M^*(E)$ satisfies (5), then, as we have shown above [equivalence of (5) and (0)], $M^*(E)$ is binary and so $M(E)$ is too by Proposition 2.1.4. Let C_1^* and C_2^* be two cocircuits in $M^*(E)$, forming a pair and such that $C_1^* \neq C_2^*$ and $C_1^* \cap C_2^* \neq \varnothing$. The hyperplanes $H_1 = E \backslash C_1^*$ and $H_2 = E \backslash C_2^*$ in $M(E)$ have thus a coline L as intersection (see appendix), and we have $H_1 \cup H_2 \neq E$. Since there is at most one other hyperplane containing L, $(E \backslash (H_1 \cup H_2)) \cup L = E \backslash (H_1 \bigtriangleup H_2)$ is a hyperplane of $M(E)$, whence, on looking back at the dual, $C_1^* \bigtriangleup C_2^*$ is a circuit in $M^*(E)$. This completes the proof of the theorem.   □

The proofs of equivalence we have given are all direct and make no appeal to other theorems concerning the representation of matroids. The general representation theorem of Tutte (Proposition 1.5.5) would, for example, have given directly the implication $(5) \Rightarrow (0)$. In the same way, the 'scum theorem' (see White 1986, Theorem 8.4.1) would give directly the equivalence of conditions (7) and (8), these conditions being interpreted in the language of lattice theory, as indicated above for condition (7).

Finally, we note that one can see directly that the matroid $U_{2,4}$, that unique smallest non-binary matroid, in the sense of condition (7), is not binary, because the projective line over the field $GF(2)$ has only three points, whereas $U_{2,4}$ has four points.

Remark. It follows from condition (1) that *a binary matroid is completely defined by the fundamental system of circuits* relative to any one basis. This is false for arbitrary matroids, as the following example illustrates: let M be the matroid of the (multi-) graph G of Figure 2.1, with edges numbered 1 to 4, and let P be the matroid $U_{2,4}$ on the set $\{1, 2, 3, 4\}$. Then M and P both admit $B = \{1, 2\}$ as basis, and have the same fundamental system of circuits with respect to B, that is, $C(3) = \{1, 2, 3\}$, $C(4) = \{1, 2, 4\}$. The matroids are, however, different, one being binary (that is, M, which is graphic, concerning which, see below), the other not, as we have already seen.

Figure 2.1. A multigraph.

2.3. Related Characterizations

Certain of the conditions characterizing binary matroids in theorem 2.2.1 can be weakened.

Condition (1) can be stated as applying to one single basis (Las Vergnas 1980). It becomes:

(1') *There is a basis B such that for every circuit C,*

$$C = \underset{X \in C \backslash B}{\triangle} C(x).$$

In the case of connected matroids, condition (4) can be stated as applying only to those circuits which contain a given element in E (Bixby 1974). This comes from the fact that a connected binary matroid is determined by those circuits which contain a given element in E. Thus if $M(E)$ *is connected*, we have the following characterization:

(4') *Let e be an element in the set E. Given any two distinct circuits C_1, C_2 not containing e, then $C_1 \triangle C_2$ contains a circuit.*

Condition (6) can be weakened as follows (Seymour 1976):

(6') *For any circuit C and any cocircuit C^*, $|C \cap C^*| \neq 3$.*

The last following characterization is stated in terms of a new property of circuits (Fournier 1974b). We say that a circuit C_1 *distinguishes* circuits C_2 and C_3 if $C_2 \backslash C_1 \neq C_3 \backslash C_1$. Then, *a matroid is binary if and only if for any three circuits C_1, C_2, C_3, such that $C_1 \cap C_2 \cap C_3 \neq \varnothing$, there is at least one of the three circuits which distinguishes the other two.*

2.4. Spaces of Circuits of Binary Matroids

When a matroid $M(E)$ is binary, the disjoint unions of circuits of $M(E)$ form, with symmetric difference as operation, a group of subsets of E [see condition (2) of Theorem 2.2.1], which is in fact a vector space, a subspace of the $GF(2)$ vector space of subsets of E. This group is the *space* (or *group*) *of circuits* of $M(E)$. It is spanned by the circuits of $M(E)$, which are the minimal elements with respect to inclusion of subsets of E. Conversely, given a set of subsets of E, which form a group with symmetric difference as operation, it is easy to verify that the minimal members of this group are the circuits of a binary matroid in E. The matroid which defines in this way the space of circuits of a given matroid $M(E)$ is the matroid $M(E)$ itself, and the space of circuits of the matroid defined by a group of subsets is the group itself.

Thus, binary matroid and group of subsets of E are two equivalent structures [the term 'group of subsets' was chosen by Ghouila-Houri (1964), who studied binary matroids in this context.]

Duality of matroids is interpreted, in the case of binary matroids, in a remarkable way, as orthogonality between a space of circuits and a space of cocircuits. To be precise, let $M(E)$ be a binary matroid, Γ the space of circuits of $M(E)$, and Γ^* the space of cocircuits of $M(E)$, that is, the space of circuits of $M^*(E)$. Then Γ and Γ^* are orthogonal with respect to the inner product in the space of subsets of E given by $\langle A, B \rangle = |A \cap B|_2$ (remainder modulo 2).

2.5. Coordinatizing Matrices of Binary Matroids

Given a matroid $M(E)$, let $B = \{e_1, \ldots, e_n\}$ be a basis of $M(E)$, and assume that $E = \{e_1, \ldots, e_n, e_{n+1}, \ldots, e_N\}$. Denote by $C(e_j)$ the *fundamental* circuit of e_j with respect to the basis B, if $j \geqslant n + 1$, and let $C(e_j) = \{e_j\}$ if $j \leqslant n$. Let $\mathcal{M} = (a_{ij})$ be the matrix with $1 \leqslant i \leqslant n$, $1 \leqslant j \leqslant N$
and

$$a_{ij} = \begin{cases} 1 & \text{if } e_i \in C(e_j) \\ 0 & \text{otherwise.} \end{cases}$$

Then condition (2) implies that $\mathcal{M}$ is the coordinatizing matrix in echelon form with respect to the basis B of $M(E)$ as discussed in Chapter 1.

2.6. Special Classes of Binary Matroids; Graphic Matroids

Another very important class of binary matroids is that of the *unimodular* (or *regular*) *matroids*, those matroids which are coordinatizable over every field. They can be represented by a totally unimodular matrix, that is, by a matrix in which the determinant of every square submatrix is 0, +1 or −1. These matroids are discussed in Chapter 3. We here simply recall that regular matroids were characterized by Tutte as those binary matroids not having as minor either the projective plane over $GF(2)$ (the *Fano matroid*) or its dual.

Among the unimodular matroids there is a special class which has been considered since the early days of matroid theory, in the work of Whitney (1935). These are the graphic and cographic matroids, which have also been examined in Chapter 6 of White (1986).

Given a graph $G = (V, E)$, it is easy to show that the elementary cycles, as subsets of the set of edges (that is, those sets of edges of G which span a connected subgraph in which every vertex has valence 2), are the *circuits* of a matroid on the set E of edges of G. This matroid is the *cycle matroid* of the graph G.

A matroid is graphic if it is isomorphic to the cycle matroid of a graph.

In the same way, the elementary cocycles of a graph G are the circuits of a matroid on E, called the *cocycle matroid* of G. This matroid is dual to the cycle matroid of G. Such matroids form the class of *cographic* matroids, the dual of the class of graphic matroids.

It is well-known that, in a graph, the symmetric difference of two elementary cycles is the disjoint union of elementary cycles. From this fact, it is clear that graphic matroids, and thus also cographic matroids, are binary. In fact, these matroids are representable over any field, that is, they are unimodular.

Characterization by excluded configurations. The following theorem of Tutte (1965, see also Ghouila Houri 1964) is one of the most remarkable results of matroid theory.

2.6.1. Theorem. *A matroid is graphic if and only if it has no minor isomorphic to the matroid $U_{2,4}$, to the Fano matroid F, to its dual F*, or to the cocycle matroid of either of the two Kuratowski graphs K_5 or $K_{3,3}$ (see Figure 2.2).*

For a new proof of this theorem, see Seymour (1979).

Figure 2.2. The two Kuratowski graphs.

$$K_5 \qquad\qquad K_{3,3}$$

Remarks. (1) The excluded minors $U_{2,4}$, F, F* characterize, as we have seen, the unimodular matroids. Thus the graphic matroids are characterized as those unimodular matroids which have no minor isomorphic to the cocycle matroids of the two graphs K_5 and $K_{3,3}$.

(2) Note that the cographic matroids are characterized by the exclusion as minors of the duals of the cocycle matroids of those two graphs, that is, by the exclusion as minors of the cycle matroids of K_5 and $K_{3,3}$.

Tutte's theorem is a generalization to matroids of the celebrated theorem of Kuratowski, which characterizes planar graphs. The generalization is by way of the no less celebrated theorem of Whitney:

2.6.2. Theorem. *A graph is planar if and only if its cocycle matroid is graphic.*

An elementary proof of this theorem which depends on Kuratowski's Theorem for graphs may be found in Theorem 6.1.7. of White (1986). By applying the condition in Tutte's theorem to the cocircuit matroid, we obtain the characterization of planar graphs by the exclusion as minors of the two Kuratowski graphs (or rather, by the exclusion of subdivisions of these two graphs).

Note moreover that, in general, every characterization of graphic matroids provides in this way a characterization of planar graphs, or generalizes such a characterization. On this subject see, among others, the papers of Welsh (1969a) and Fournier (1974a, 1974b).

The main significance of Whitney's theorem has to do with the fact that duality of matroid parallels duality of planar graphs, and, as Whitney observed, even if a graph does not always have a dual, its cycle matroid always does. And when this dual of the cycle matroid is graphic, under certain conditions an associated graph is a planar dual of the given graph.

Remark. The Euler relation for planar graphs is nothing but the expression for the equality between the ranks of two isomorphic matroids: the cycle matroid M of the given graph, and the cocycle matroid M' of a planar dual. In detail if G' is a planar dual of G, denoting by n and n' the numbers of vertices of the graphs G, G', respectively, by m the common number of edges of G and G', the rank of M is equal to $n-1$, that of M' is equal to $m-(n'-1)$, and since n' is also the number of faces of the planar representation of G (dual to that of G'), we arrive at the Euler relation

$$n - m + f = 2.$$

We could continue to give many properties of graphic matroids, for instance to describe their independent sets and bases (which correspond to maximal trees in the graph), or to describe their circuit and cocircuit spaces (which correspond to the spaces of cycles and cocycles, elementary or not, in the graphs), or to make precise the duality relation between them. For all this, we refer the reader to the specialised references, and to Chapter 6 of White (1986).

2.7 Appendix on Modular Pairs of Circuits in a Matroid

Two circuits C_1 and C_2 in a matroid $M(E)$, in which r is the rank function, form a *modular pair* if they satisfy the modular relation:

$$r(C_1) + r(C_2) = r(C_1 \cup C_2) + r(C_1 \cap C_2).$$

This relation implies, when $C_1 \neq C_2$, that

$$r(C_1 \cup C_2) = |C_1 \cup C_2| - 2.$$

We then verify easily that given two *distinct* circuits C_1 and C_2 in $M(E)$, and looking at the hyperplanes $H_1^* = E \backslash C_1, H_2^* = E \backslash C_2$ in $M^*(E)$, we have:

C_1 and C_2 form a modular pair if and only if $H_1^* \cap H_2^*$ is a coline in $M^*(E)$.

2.7.1. Lemma. *(White 1971) Let C_1 and C_2 be a modular pair of circuits such that $C_1 \neq C_2, C_1 \cap C_2 \neq \varnothing$, and let a be an element of $C_1 \cap C_2$. There exists a unique circuit C_3 such that*

$$C_3 \subseteq (C_1 \cup C_2) \backslash \{a\},$$

and furthermore,

$$C_3 \supset C_1 \triangle C_2.$$

Proof. The existence of C_3 is easier to see on passing to the dual: the modularity of C_1 and C_2 implies that of $H_1^* = E - C_1$ and $H_2^* = E - C_2$; there is one and only one hyperplane H_3^* containing the coline $H_1^* \cap H_2^*$ and the element a; this is the hyperplane spanned by $(H_1^* \cap H_2^*) \cup \{a\}$, and, furthermore, $H_3^* \cap (H_1^* \triangle H_2^*) = \varnothing$ because $H_3^* \cap H_1^* = H_3^* \cap H_2^* = H_1^* \cap H_2^*$. So $C_3 = E \backslash H_3^*$ satisfies the stated conditions. $\qquad\square$

2.7.2. Lemma. *Given a basis B and a circuit C such that $|C \backslash B| \geq 2$, there is a modular pair C_1, C_2 of circuits such that $C_1 \neq C_2$, $C_1 \cap C_2 \neq \emptyset$ and $C_1 \triangle C_2 \subseteq C$.*

Proof. Let $H^* = E \backslash C$, a hyperplane in $M^*(E)$. There is a coline L^* in $M^*(E)$ such that $E \backslash (B \cup C) \subset L^* \subset H^*$ (otherwise one would have $H^* = E \backslash (B \cup C)$, and $B \subset C$, which is impossible). Let e_1 be an element in $C \backslash B$, let H_1^* be the hyperplane spanned in $M^*(e)$ by $L^* \cup \{e_1\}$, and choose an element e_2 in $C \backslash (B \cup H_1^*)$. This set $C \backslash (B \cup H_1^*)$ is non-empty because otherwise the hyperplane H_1^* in $M(E)$ would contain the basis $E \backslash B$. Finally, let H_2^* be the hyperplane spanned in $M^*(E)$ by $L^* \cup \{e_2\}$. Then the circuits $C_1 = E \backslash H_1^*$ and $C_2 = E \backslash H_2^*$ are distinct (as are H_1^* and H_2^*), they form a modular pair (because $H_1^* \cap H_2^*$ is a coline), and they satisfy the condition $C_1 \cap C_2 \neq \emptyset$ (because $H_1^* \cup H_2^* \neq E$, there being elements of H^* outside of $H_1^* \cup H_2^*$) and $C \supseteq C_1 \triangle C_2$ [because $H^* \cap (H_1^* \triangle H_2^*) = \emptyset$]. $\qquad\square$

Exercises

2.1. Give a direct proof of Propositions 2.1.3 and 2.1.4, without using general theorems about coordinatizable matroids.

2.2. For what values of r and n is the matroid $U_{r,n}$ binary?

2.3. Given a matroid M on a set E, its collection $\mathscr{I}$ of independent subsets, and an integer k less than or equal to the rank of M, the *rank k truncation* of M is the matroid on E that has as its family of independent sets the set

$$\mathscr{I}_k = \{L | L \in \mathscr{I} \text{ and } |L| \leq k\}.$$

Show that the truncation of a binary matroid is not, in general, binary.

2.4. A matroid M on a set E is said to be *Eulerian* if E can be expressed as the disjoint union of circuits of M. M is said to be *bipartite* if every circuit of M has even cardinality. Show that a binary matroid is Eulerian if and only if its dual is bipartite, (Welsh 1969b).

2.5. Let V be a finite dimensional $GF(2)$-vector space of functions from a finite set E into $GF(2)$. Let $f | X$ denote the restriction of the function f to a subset X. We define

for each subset $X \subseteq E$, $h(X) = \{f \in V | f | X = 0\}$, and

for each subset $U \subseteq V$, $k(U) = \{x \in E | f(x) = 0, \forall f \in U\}$.

Show that the mapping $\psi = k \cdot h$ is the dependence closure of a binary matroid on E. Study the converse. Deduce a characterization of binary matroids in terms of binary function spaces.

References

Bixby, R.E. (1974). *l*-matrices and a characterization of binary matroids. *Discrete Math.* **8**, 139–45.

Fournier, J.-C. (1974a). Matroides graphiques et graphes topologiques. Thesis, Université Paris 6.

Fournier, J.-C. (1974b). Une relation de séparation entre cocircuits d'un matroide. *J. Comb. Theory Ser. B* **16**, 181–90.

Ghouila-Houri, A. (1964). Flots et tensions dans un graphe. *Ann. Scient. Ec. Norm. Sup.* **81**, 267–339.

Las Vergnas, M. (1980). Fundamental circuits and a characterization of binary matroids. *Discrete Math.* **31**, 327.

Lehman, A. (1964). A solution of the Shannon switching game. *J. Soc. Indust. Appl. Math.* **12**, 687–725.

Minty, G.J. (1966). On the axiomatic foundations of the theories of directed linear graphs....*J. Math. Mech.* **15**, 485–520.

Rado, R. (1957). A note on independence functions. *Proc. Lond. Math. Soc.* **7**, 300–20.

Seymour, P.D. (1976). The forbidden minors of binary clutters. *J. London Math. Soc.* (2) **12**, 356–60.

Seymour, P.D. (1979). On Tutte's characterization of graphics matroids. (*Proc. Colloq. Univ. Montreal, Montreal 1979*) Part I *Ann. Discrete Math.* **8**. (1980), 83–90.

Tutte, W.T. (1965). Lectures on matroids. *J. Res. Nat. Bur. Stand.* **69B**, 1–47.

Welsh, D.J.A. (1969a). On the hyperplanes of a matroid. *Proc. Cambridge Phil. Soc.* **65**, 11–18.

Welsh, D.J.A. (1969b). Euler and bipartite matroids. *J. Comb. Theory*, **6**, 375–7.

White, N. (1971). *The Bracket ring and combinatorial geometry*. Thesis, Harvard University,

White, N., ed. (1986). *Theory of Matroids*, Cambridge University Press.

Whitney H. (1933). Planars graphs. *Fund. Math* **21**, 73–84.

Whitney H. (1935). On the abstract properties of linear dependence. *Ann. J. Math* **57**, 509–33.

3

Unimodular Matroids

NEIL WHITE

3.1. Equivalent Conditions for Unimodularity

Unimodular matroids were defined in Chapter 1 as the class of matroids which may be coordinatized over every field. In Theorem 3.1.1 we give a number of equivalent characterizations of this class. Certainly the two most striking and powerful of these are Tutte's excluded minor characterization and Seymour's decomposition [conditions (8) and (9) of Theorem 3.1.1]. We first need some definitions and notation.

A coordinatization of $M(S)$ over $\mathbb{Q}$ given by $n \times N$ matrix A with integer entries, and $n < N$, is said to be *totally unimodular* if every $k \times k$ submatrix has determinant equal to 0 or ± 1, for all k, $1 \leqslant k \leqslant n$, and is said to be *locally unimodular* if every $n \times n$ submatrix has determinant equal to 0 or ± 1.

Let D be the bond-element incidence matrix of $M(S)$. That is, if $R_1, R_2, \ldots, R_m$ are the bonds of M and $S = \{x_1, x_2, \ldots, x_N\}$, then $D = (b_{ij})$, with $b_{ij} = 1$ if $x_j \in R_i$, and $b_{ij} = 0$ otherwise. Similarly, let E be the circuit-element incidence matrix of M. Suppose that it is possible to change some of the entries of D from 1 to -1 to get a matrix D', and similarly, change E to E', so that $D'(E')^t = 0$ over $\mathbb{Q}$ (where t denotes transpose). Then we say that M is *signable*. [This is closely related to the notion of orientability, considered in a chapter of White (1988).]

In Section 7.6 of White (1986) 1-sums, 2-sums, or (for binary matroids) 3-sums of two matroids $M_1(E_1)$ and $M_2(E_2)$ were defined as $P_x(M_1, M_2) - x$, where $P_x(M_1, M_2)$ is the generalized parallel connection across a flat x, and x is empty, a point, or a 3-point line (respectively). To avoid triviality we insist that $P_x(M_1, M_2) - x$ have larger cardinality than M_1 or M_2. For binary matroids, with which we are concerned here, an equivalent definition is to say that each of these three sums is the matroid $M_1 \triangle M_2$ on the symmetric difference $E_1 \triangle E_2$ which has as its cycles (i.e., disjoint unions of circuits) all subsets of the form $C_1 \triangle C_2$, where C_i is a cycle of M_i. Then

(A) $M_1 \triangle M_2$ is the 1-sum of M_1 and M_2 if $E_1 \cap E_2 = \varnothing$ and $E_1 \neq \varnothing$, $E_2 \neq \varnothing$.

(B) $M_1 \triangle M_2$ is the 2-sum of M_1 and M_2 if $E_1 \cap E_2 = \{e\}$, e is neither a loop nor an isthmus of M_1 or of M_2, and $|E_1| \geqslant 3$, $|E_2| \geqslant 3$.

(C) $M_1 \triangle M_2$ is the 3-sum of M_1 and M_2 if $E_1 \cap E_2 = L$, where $|L| = 3$, L is a line (and therefore L is a circuit) in each of M_1 and M_2, L includes no bond of M_1 or M_2, and $|E_1| \geqslant 7$, $|E_2| \geqslant 7$.

In fact, the 1-sum is just direct sum. The 2-sum is just pasting together of M_1 and M_2 at the common element e, followed by the deletion of e, so that the rank of $M_1 \triangle M_2$ is as large as possible, namely $rM_1 + rM_2 - 1$. The 3-sum is a similar pasting together along a common line, again keeping the rank as large as possible, namely $rM_1 + rM_2 - 2$.

The matroid R_{10} in the following theorem is given in Exercise 1.7. $U_{2,4}$ is the 4-point line, F_7 the 7-point Fano plane, and F_7^* the orthogonal dual of F_7.

A matroid is called *unimodular* (or *regular*) if it satisfies any of the conditions of the following theorem.

3.1.1. Theorem. *The following conditions are equivalent, for a matroid $M(S)$.*

(1) *M has a totally unimodular coordinatization over $\mathbb{Q}$.*

(2) *M has a locally unimodular coordinatization over $\mathbb{Q}$.*

(3) *The brackets for M may be assigned the values $0, \pm 1$ in $\mathbb{Q}$ so that the syzygies of Proposition 1.6.1 are satisfied.*

(4) *M may be coordinatized over K, for every field K.*

(5) *M may be coordinatized over $GF(2)$ and over K, for some K with char $K \neq 2$.*

(6) *M is signable.*

(7) *For every hyperplane H of M there exists a function $F_H : S \to \mathbb{Q}$ such that kernel $F_H = H$ for every H, image $F_H \subseteq \{0, 1, -1\}$, and for every three hyperplanes H_1, H_2, and H_3 containing a common coline, there exists α_1, α_2, and $\alpha_3 \in \{1, -1\}$ such that $\alpha_1 F_{H_1} + \alpha_2 F_{H_2} + \alpha_3 F_{H_3} = 0$.*

(8) *M has no minor isomorphic to $U_{2,4}$, F_7, or F_7^*.*

(9) *M may be constructed by 1-, 2-, and 3-sums from graphic matroids, cographic matroids, and matroids isomorphic to R_{10}.*

Proof of the equivalent of (1) through (5). $(1) \Rightarrow (2)$ and $(4) \Rightarrow (5)$ are trivial, and $(2) \Rightarrow (3) \Rightarrow (4)$ are immediate from Proposition 1.6.1, where the bracket values $0, \pm 1 \in \mathbb{Q}$ are simply regarded as elements of the field K. Since the syzygies hold over $\mathbb{Q}$, they also hold mod p, where $p = \text{char } K$.

We now have only to prove $(5) \Rightarrow (1)$. This proof is due to Brylawski (1975). Let $A = (I_n | L)$ be a coordinatization of $M(S)$ over K, where $M(S)$ is binary and char $K \neq 2$. We assume that A is in (B, T)-canonical form, where T is a spanning tree of the bipartite graph Γ whose adjacency matrix is determined by L (see Section 1.2). We now claim that each entry in L (and hence in A) is 0 or ± 1. Let w be a non-zero entry of L, other than one of the entries corresponding to T. Then w corresponds to an edge of $\Gamma - T$, and hence has a basic circuit C in Γ. We will prove that $w = \pm 1$ by induction on the size of C.

It is not difficult to see that the edges of C correspond to a cyclic sequence of $2k$ non-zero entries of L, for some $k \geqslant 2$, with the property that each odd-numbered entry in the sequence is in the same column as its predecessor, and each even-numbered entry in the same row as its predecessor. For example, the submatrix containing the sequence of entries may look like the following:

$$\begin{matrix} 0 & 1 & 1 & 0 \\ 1 & 1 & 0 & 0 \\ 1 & 0 & 0 & 1 \\ 0 & 0 & 1 & w \end{matrix}$$

Now either these $2k$ entries are the only non-zero entries in a $k \times k$ submatrix of L, or if there are other entries, w forms a circuit of size less than $2k$ with entries which are either in T or themselves have basic circuits of size $< 2k$. By the induction hypothesis, these other entries are all ± 1, hence in any case we get a $j \times j$ submatrix J of L having exactly $2j$ non-zero entries, with 2 in each row and column, and each entry except w equal to ± 1. But then J is uniquely the sum of two permutation matrices, so $\det J = \pm 1 \pm w$. But since $M(S)$ is binary, it may also be coordinatized over $GF(2)$ by replacing each non-zero entry in A by 1 in $GF(2)$, since basic circuits of M must be preserved. But then over $GF(2)$, $\det J = 0$, hence we must also have $\det J = 0$ over K to preserve dependence. Therefore $w = \pm 1$.

The proof of (1)–(5) will now be complete if we prove the following lemma, by regarding A as a matrix over $\mathbb{Q}$, since the operations in the proof of the lemma do not depend on the characteristic.

3.1.2. Lemma. *Let $M(S)$ be a binary matroid which is coordinatized by a matrix A in echelon form over $\mathbb{Q}$, with every entry of A equal to 0 or ± 1. Then A is totally unimodular.*

Proof. Let W be a square submatrix of A. We now do row operations on W to reduce it to echelon form. Given $w_{ij} \neq 0$, since $w_{ij} = \pm 1$, we add $-w_{ij}w_{hj}$ times row i to row h for each h, to get w_{ij} to be the only non-zero entry in column j. Now consider an entry w_{hk} in the original submatrix W, where $h \neq i, k \neq j$. Then the above row operations replace w_{hk} by $w_{hk} - w_{ij}w_{hj}w_{ik}$, which is 0 or ± 1 unless $w_{hk} = -w_{ij}w_{hj}w_{ik} \neq 0$. But then the following 2×4 submatrix existed in the matrix A:

$$\begin{bmatrix} 1 & 0 & w_{ij} & w_{ik} \\ 0 & 1 & w_{hj} & w_{hk} \end{bmatrix}.$$

This submatrix coordinatizes a minor of $M(S)$ which is isomorphic to L_4, a contradiction to the assumption that $M(S)$ is binary.

Thus the reduction of W to echelon form may be completed while keeping all entries 0 or ± 1. Thus det $W = 0$ or ± 1, and A is totally unimodular

$\square$

Proof of equivalence of (1) *through* (7). First we will show $(1) \Rightarrow (6) \Rightarrow (5)$.

Let A be a totally unimodular matrix over $\mathbb{Q}$ coordinatizing $M(S)$ and in echelon form with respect to the basis B. Then the i-th row of A is non-zero on precisely the elements of the basic bond $S - \overline{B - \{b_i\}}$ corresponding to the i-th element of B. Furthermore, by row operations, we may bring A into echelon form A' with respect to any other basis B', thus obtaining a row for any bond of M. Furthermore, A' must still be totally unimodular, since $n \times n$ determinants are preserved by the row operations, and any $k \times k$ determinant of A' may be augmented by columns from B' to obtain an $n \times n$ determinant, at most changing the sign of the determinant.

Now let D' be a matrix obtained by taking such a row for each bond of M. D' is then just the bond-element incidence matrix with some 1's changed to -1's, and the row-space of D' is the same as the row-space of A. By Proposition 1.3.1, $M^*(S)$ also has a coordinatization A^* obtained from A by transposing. It is very easy to check that A^* is also totally unimodular. Letting E' be the matrix obtained for the bonds of M^* as D' was for M, we see that E' is just the circuit-element incidence matrix of M with some 1's changed to -1's. Furthermore, the rows of D' and the rows of E' are orthogonal, again by Proposition 1.3.1, hence $D'(E')^t = 0$, proving $(1) \Rightarrow (6)$.

Now suppose that we are given D' and E' as above, with $D'(E')^t = 0$. Since each row of D' is orthogonal to each row of E', we see immediately that if R is a bond and C a circuit of $M(S)$, then $|R \cap C|$ is even. Thus from Theorem 2.2.1, M is binary.

Let B be a basis of $M(S)$ and assume the elements of S have been ordered so that the elements of B come first. The basic bonds $S - \overline{B - \{b_i\}}$ for $b_i \in B$ give us a submatrix D'' of D',

$$D'' = (I' | U),$$

where I' is the matrix of columns corresponding to the elements of B', and I' is an $n \times n$ identity matrix with some of the entries possibly changed from 1 to -1. Now, the dimension of the row-space of D' is at least n, the dimension of the row-space of D''.

Similarly, by taking the rows of E' corresponding to the basic circuits of B, we have

$$E'' = (V | I''),$$

where I'' is the matrix of columns corresponding to $S - B$, and I'' is an $(N - n) \times (N - n)$ identity matrix with some of the entries possibly changed from

1 to -1. The dimension of the row-space of E' is at least $N - n$, the dimension of the row-space of E''. Since the row-spaces of D' and E' are orthogonal subspaces of an N-dimensional vector-space over $\mathbb{Q}$, we have equality in both cases, that is, row-rank $(D') = n$ and row-rank $(E') = N - n$.

We will now show that D'' is a totally unimodular matrix coordinatizing $M(S)$. Let B' be any basis of M. If we construct D''' from the basic bonds of B' as we did D'' from B (but keeping the ordering of the elements of S fixed), we see that D''' and D'' are row-equivalent, since the rows of each are a basis of the row space of D'. Thus the columns of D'' corresponding to B' are linearly independent.

Let C be a circuit of $M(S)$. Then C corresponds to a row e_c of E' which is orthogonal to the rows of D'', and hence the entries of e_c are the coefficients of a linear dependence of the columns of D'' corresponding to the elements of C. Thus D'' is a coordinatization of M, and by Lemma 3.1.2 it is also unimodular.

Thus (1)–(6) are equivalent. The equivalence of these with (7) now follows easily by noting that the functions f_H correspond to rows of the signed bond-element matrix D', with f_H in particular corresponding to the row for the bond $S - H$.

Proof of conditions (8) and (9). The implication (5)$\Rightarrow$(8) is easy, since L_4 cannot be a minor if M is binary, and F_7 or F_7^* cannot be coordinatized over any field whose characteristic is not 2 (see Exercise 1.9). The converse was proved by Tutte using his very deep Homotopy Theorem (Tutte 1958), and is certainly one of the most beautiful and important results in matroid theory. We state the Homotopy Theorem and sketch the proof of (8)$\Rightarrow$(7) in the next section.

The implication (9)$\Rightarrow$(5) is easy by observing that 1-sums, 2-sums, and 3-sums preserve coordinatizability over $GF(2)$ and $GF(3)$ [see p. 186 of White (1986)]. Seymour's Theorem (1980) is (8)$\Rightarrow$(9). The proof is much too long to be included here. One advantage of this result is that it includes Tutte's Theorem as a corollary. $\qquad\square$

3.2 Tutte's Homotopy Theorem and Excluded Minor Characterization

We now give a careful statement of Tutte's Homotopy Theorem, and sketch its use to prove Tutte's excluded minor characterization of unimodular matroids. There are several reasons why we choose to do so. The first is the historical importance of Tutte's work, despite the fact that his excluded minor characterization can now also be proved by Seymour's method. The second is the importance of the ideas involved for further work in coordinatizations. This importance seems restricted by Tutte's heavy use of the crucial property of binary matroids that coline is contained in at most three distinct hyperplanes (or copoints). Nevertheless, both Reid (unpublished) and Bixby (1979) were able to extend Tutte's methods to obtain the excluded minor characterization of ternary matroids. The third reason is that such a sketch of Tutte's ideas is not available in accessible form elsewhere, except in Tutte's own writing. Although Tutte's terminology and notation are perhaps suitable for someone who is interested primarily in the graph-theoretical aspects of matroid theory, they are quite confusing to the large majority of matroid theorists who use terminology similar to that used in these volumes. For example, what Tutte calls a point is in our terminology a bond, and for our purposes is best complemented to get a hyperplane. It is hoped that the translation provided here will be useful not only as an overview of Tutte's methods, but also as an entry point to Tutte's papers for those who wish to study them in detail.

We first need some definitions. A *copoint* (or hyperplane), *coline*, or *coplane* in a matroid $M(E)$ of rank n is a flat of rank $n-1, n-2$, or $n-3$ (respectively). A flat Y is *T-connected* if $M(E)/Y$ is connected. A *path* in M is a sequence $(X_1, X_2, \ldots, X_k)$ of copoints such that for $1 \leqslant i \leqslant k-1$, $X_i \cap X_{i+1}$ is a T-connected coline. Thus each such coline $X_i \cap X_{i+1}$ is contained in a third copoint distinct from X_i and X_{i+1}. A collection $\mathscr{C}$ of copoints of M is a *linear subclass of copoints* (see White 1986, Exercise 7.8) if whenever X_1, X_2, and X_3 are distinct copoints all containing a common coline, and $X_1 \in \mathscr{C}$ and $X_2 \in \mathscr{C}$, then $X_3 \in \mathscr{C}$. A path is *off* $\mathscr{C}$ if no copoint of the path is a member of $\mathscr{C}$. A path is *closed* if the first and last copoints in the path are identical. We now describe four types of closed paths which will be called *elementary paths off* $\mathscr{C}$, for a particular linear subclass $\mathscr{C}$.

(1) (X, Y, X), an arbitrary closed path of length 2 off $\mathscr{C}$.
(2) (X, Y, Z, X), a closed path of length 3 off $\mathscr{C}$ such that $X \cap Y \cap Z$ is either a coline or a coplane.
(3) (X, Y, Z, T, X), a closed path of four distinct copoints off $\mathscr{C}$, where $X \cap Y \cap Z \cap T$ is a coplane P, $X \cap Y$ and $Z \cap T$ span a copoint A, $X \cap T$ and $Y \cap Z$ span a copoint B, $A \in \mathscr{C}$, $B \in \mathscr{C}$, and every T-connected coline containing P is contained either in A or in B.
(4) (A, X, B, Y, A), a closed path of four distinct copoints off $\mathscr{C}$ where $A \cap X \cap B \cap Y = D$ and the contraction $M(E)/D$ is a matroid of rank 4 containing six distinct points $P_1, P_2, \ldots, P_6$ with A/D spanned by $\{P_2, P_3, P_5, P_6\}$, B/D by $\{P_1, P_3, P_4, P_6\}$, X/D by $\{P_2, P_3, P_4\}$, Y/D by $\{P_1, P_2, P_6\}$, and with $\{P_1, P_2, P_4, P_5\}$ spanning another copoint off $\mathscr{C}/D$, where $\mathscr{C}/D = \{X/D : X \in \mathscr{C}\}$. Furthermore, $\{P_1, P_2, P_3\}$, $\{P_1, P_5, P_6\}$, $\{P_2, P_4, P_6\}$, and $\{P_3, P_4, P_5\}$ all span copoints which are in $\mathscr{C}/D$, and all other points of M/D are on the three lines P_1P_4, P_2P_5, and P_3P_6.

Now, if $P = (X_1, X_2, \ldots, X_k)$ and $R = (X_k, X_{k+1}, \ldots, X_m)$ are two paths, we define their product PR as the path $(X_1, X_3, \ldots, X_k, \ldots X_m)$. If $Q = (X_k, \ldots, X_k)$ is one of the elementary paths defined above, we say that PQR and PR are *elementary deformations* of each other with respect to $\mathscr{C}$. Two paths P and P' off $\mathscr{C}$ are *homotopic* with respect to $\mathscr{C}$ if one may be obtained from the other by a finite sequence of elementary deformations with respect to $\mathscr{C}$. Homotopy is clearly an equivalence relation.

3.2.1 Proposition. *Let $\mathscr{C}$ be a linear subclass of copoints in a connected matroid $M(E)$, and let X and Y be copoints of M such that $Y \notin \mathscr{C}$. Then there exists a path from X to Y which is off $\mathscr{C}$ with the possible exception of the first copoint X.*

A proof of this in our notation may be found in Crapo & Rota (1970).

3.2.2. Proposition. (*Tutte's Homotopy Theorem*). *Let $\mathscr{C}$ be any linear subclass of the matroid $M(E)$, and let P be any closed path off $\mathscr{C}$. Then P is homotopic to a trivial path with respect to $\mathscr{C}$.*

We omit the proof of Proposition 3.2.2 since it is fairly long and technical. We prefer instead to show how it is applied to prove the excluded minor characterization for unimodular matroids.

3.2.3. Theorem. *A matroid M is unimodular if and only if M is binary and has no minor isomorphic to the Fano plane F_7 or the orthogonal matroid F_7^*.*

Proof. We have already observed that the necessity is easy. To prove the sufficiency, suppose that M is a minimal matroid such that M is binary with no minor isomorphic to F_7 or F_7^* and yet M is not unimodular. Then for arbitrary $a \in E, M - a = M'$ is unimodular. Let $\mathscr{C}$ be the linear subclass of copoints X of M' such that $a \in \mathrm{cl}(X)$ in M.

Now we fix a unimodular coordinatization of M', given by $f_X : E \to \{0, \pm 1\} \subseteq \mathbb{Q}$ for every copoint X of M', as in Proposition 1.5.5. Our task is to construct such an f_X for every copoint X of M.

Let X and Y be copoints of M' on a T-connected coline, with X and Y off $\mathscr{C}$. Then there exists $x \in E - (X \cup Y \cup \{a\})$. Let $t(X, Y) = f_X(x) f_Y(x)$. Then $t(X, Y)$ is independent of the choice of x, for if $y \in E - (X \cup Y \cup \{a\})$ and $f_X(x) f_Y(x) \neq f_X(y) f_Y(y)$, then the coordinatizing matrix can easily be shown to have a submatrix

$$
\begin{array}{cccc}
1 & 0 & f_X(x) & f_X(y) \\
0 & 1 & f_Y(x) & f_Y(y)
\end{array}
$$

which implies a minor L_4 of M', a contradiction.

Now let $R = (X_1, X_2, \ldots, X_k)$ be any path in M' off $\mathscr{C}$. We define $u(R) = \prod_{i=1}^{k-1} t(X_i, X_{i+1}) = \pm 1$, and claim that $u(R) = 1$ for every closed path off $\mathscr{C}$. To prove this claim, it suffices by the Homotopy Theorem to prove that $u(R) = 1$ for each of the four elementary paths off $\mathscr{C}$.

(1) Let $R = (X, Y, X)$, then $u(R) = t(X, Y)^2 = 1$.

(2) Let $R = (X, Y, Z, X)$. Then X, Y, Z cannot contain a common coline L, since none of them contains the point a, and the binary matroid M cannot have four copoints on L. Therefore X, Y, and Z intersect in a coplane P. If there is a point $x \notin X \cup Y \cup Z$, then $u(R) = f_X(x) f_Y(x) f_Y(x) f_Z(x) f_Z(x) f_X(x) = 1$. If there is no such point x, then for R to be a path, we must have $e \in X - (Y \cup Z), f \in Y - (X \cup Z), g \in Z - (X \cup Y)$. Since $Y \cap Z$ is a coline, there must also be $b \in (Y \cap Z) - X$, and similarly $c \in (X \cap Z) - Y, d \in (X \cap Y) - Z$. Then these six points together with the point a induce a Fano configuration in M/P, a contradiction.

(3) In this case, we have f_X, f_Y, f_Z, f_T and it is easy to see from Lemma 1.5.6 that these four functionals are linearly dependent, since $X \cap Y \cap Z \cap T$ is a coplane P. This dependence implies that the following determinant is zero, where $b \in (X \cap Y) - P$, $c \in (Y \cap Z) - P$, $d \in (Z \cap T) - P$, $e \in (T \cap X) - P$:

$$\begin{vmatrix} 0 & 0 & f_Z(b) & f_T(b) \\ f_X(c) & 0 & 0 & f_T(c) \\ f_X(d) & f_Y(d) & 0 & 0 \\ 0 & f_Y(e) & f_Z(e) & 0 \end{vmatrix} = 0$$

which implies $u(R) = 1$.

(4) This case leads directly to F_7^* when we include the point a and contract by D, again a contradiction.

Now we are ready to construct the coordinatization of M, by defining f_X for every copoint X of M. For each copoint X, either

(A) $a \notin X$ and X is a copoint of M' (with $X \notin \mathscr{C}$),

(B) $a \in X$ and $X - a$ is a copoint of M' (with $X - a \in \mathscr{C}$), or

(C) $a \in X$ and $X - a$ is a coline of M'.

In cases (A) and (B), we already have f_X defined on $E - \{a\}$. We fix a copoint X_0 satisfying case (A), and set $f_{X_0}(a) = 1$. Then for every copoint X in case (A), there must be a path R in M' from X_0 to X off $\mathscr{C}$, by Proposition 3.2.1. Let $f_X(a) = u(R)$. Since we have already shown that $u(R) = 1$ when R is a closed path, we see that $f_X(a)$ is well-defined.

In case (B), set $f_X(a) = 0$. In case (C), $X - a = L$ must be a disconnected coline of M' (since M is binary), that is, there are copoints Y and Z of M' containing L, with $E = Y \cup Z \cup \{a\}$. Simply define $f_X = f_Y \pm f_Z$, choosing the coefficient of f_Z so that $f_X(a) = 0$.

To complete the proof, we need to show that for every three copoints X, Y, Z on a coline L, f_X, f_Y, and f_Z are linearly dependent. Suppose first that $a \notin L$. Then $a \in X$, without loss of generality. If $X - a = L$, then X is of type (C) above, and by the construction of f_X, we have the required linear dependence. If $X - a \supsetneq L$, then there exists $b \in X - L, b \neq a$. In M', we have $a f_{X-a} +$

$\beta f_Y + \gamma f_Z = 0$. Since $f_X(a) = f_X(b) = 0$ by case (B), and $f_Y(a)f_Z(a) = u(R) = t(Y, Z)$ follows from case (A) using the path $R = (Y, Z)$ off $\mathscr{C}$, and since $t(Y, Z) = f_Y(b)f_Z(b)$, we have that $\alpha f_X(a) + \beta f_X(a) + \gamma f_Z(a) = \pm(\beta f_Y(b) + \gamma f_Z(b)) = 0$, hence $\alpha f_X + \beta f_Y + \gamma f_Z = 0$.

The remaining case is $a \in L$. If $L - a$ is still a coline in M', then f_X, f_Y, and f_Z are dependent on $E - \{a\}$, and take the value zero on a, hence are dependent on E. If $L - a$ is a coplane, it is necessary to construct some additional copoints and use dependences among their f's to deduce the desired dependency. We omit the details, which are in Tutte (1958).   □

3.3. Applications of Unimodularity

An important application of Seymour's characterization of unimodular matroids [condition (9) in Theorem 3.1.1] is a polynomial algorithm for recognizing whether a matrix is totally unimodular, or more generally, whether an arbitrary matroid M is unimodular. In the general case, the number of independent sets in the matroid may be exponential compared to the rank and cardinality of the matroid, so for the problem to make sense we must assume that M is given by an *independence oracle*, a 'black box' that tells us in one step whether a given subset is independent in M. In the case of a vector matroid, for example, the independence oracle is simply a subroutine for checking linear independence. The algorithm proceeds roughly as follows:

3.3.1. Algorithm.
(1) *Check for decompositions into 1-sums, 2-sums, or 3-sums, using algorithms by Bixby and Cunningham (1981) and Cunningham and Edmonds (unpublished) for k-separations.*
(2) *Taking indecomposable matroids resulting from (1), check for graphicness by Bixby and Cunningham (1980), for cographicness by taking the orthogonal dual and checking for graphicness, and for isomorphism with R_{10}.*

This algorithm may be modified to check whether a given matrix A is unimodular as follows:

3.3.2. Algorithm.
(1) *Check that all entries of A are 0, ± 1.*
(2) *Letting M be the binary matroid on the columns of A_1, the binary matrix obtained by changing -1's to 1's in A, apply Algorithm 3.31 to determine whether M is unimodular (where we note that Algorithm 3.3.1 is easier to implement for binary matroids).*
(3) *If M is unimodular, determine a unimodular signing A_2 of A_1 (which may be determined from such signings of the graphic, cographic, and R_{10} pieces, which are easy to sign).*

(4) *Applying Proposition 1.2.5, we check whether A_2 is projectively equivalent to A, using only scalar multiplications of ± 1.*

A second application of unimodularity is in linear programming.

3.3.3. Proposition. (*Heller 1957*). *The linear program*

$$maximize \; c^t x$$
$$subject \; to \; Ax \leqslant b, \; x \geqslant 0$$

has a solution x with integer coordinates, for every choice of a vector b with integer coordinates, if and only if A is totally unimodular.

In fact many of the most efficiently solved combinatorial optimization problems, such as matroid intersection and bipartite matching, may be realized as unimodular programming problems. Indeed, this proposition makes the distinction between integer programming and linear programming no longer an issue for such problems.

There is a polynomial algorithm for solving unimodular programming problems, according to Bland and Edmonds (unpublished); see Bixby and Cunningham (1980). This algorithm uses the Seymour decomposition to reduce to the case that A is graphic or cographic. However, this case is essentially a network flow problem or its dual. One might regard this algorithm to be of no interest because of the recent highly publicized polynomial algorithms for the general linear programming problem. However, network flow problems are so efficiently solved that one can still hope for more efficient algorithms for the unimodular case than the general one.

As a third application, we consider the integer max-flow-min-cut property. This is a well-known property of directed graphs (networks), but Seymour (1977) has characterized an interesting generalization to matroids. A special element e of $M(E)$ is singled out (corresponding to an auxiliary edge from sink to source in the network case). A capacity is assigned to each element of $M(E)$ $- e$ and a flow is an assignment of a scalar to each circuit of M, such that the flow summed over all circuits containing an element x does not exceed the capacity of x. Then M has the integer max-flow-min-cut property if for every choice of e and an integer-valued capacity, there exists a non-negative integer-valued flow whose total value at e equals the minimum capacity of a cocircuit ('cut-set') of M containing e. Gallai (1959) and Minty (1966) proved independently that unimodular matroids have this property. However, Seymour (1977) has completely characterized the connected matroids with this property: they are the binary matroids with no minor isomorphic to F_7^*. This class of matroids is dual to that denoted by $\mathscr{R}'$ in Table 7.1 of White (1986). Thus they are either unimodular or contain an F_7 minor. But more is true. Matroids in $\mathscr{R}'$ must always be 2-sums of unimodular matroids and copies of F_7. This remarkable fact is an example of Seymour's concept of a *splitter*: a matroid N belonging to a hereditary class $\mathscr{F}$ which fits so tightly in $\mathscr{F}$ that any matroid M in $\mathscr{F}$ having N as a proper minor has a 1-sum or 2-sum decomposition. Thus any matroid in $\mathscr{F}$ is composed by 1-sums and 2-sums from copies of N and matroids in $\mathscr{F}$ having no minor isomorphic to N. This concept plays an important role in Seymour's proof of his characterization of unimodular matroids, in that R_{10} is a splitter for the class of unimodular matroids. Thus a stronger version of Seymour's theorem may be stated: a unimodular matroid may always be realized by 1-sums and 2-sums of copies of R_{10} and additional matroids which are 1-sums, 2-sums, and 3-sums of graphic and cographic matroids.

Finally, we mention one more application of unimodular matroids, namely, the characterization of zonotopes which pack n-dimensional Euclidean space $\mathbb{E}^n$. Let $\mathscr{S} = \{\mathbf{x}_1, \mathbf{x}_2, \ldots, \mathbf{x}_q\}$ be a set of vectors in $\mathbb{E}^n$. Without loss of generality we may assume that these vectors are non-zero and distinct up to scalar multiple, that is, that the vector matroid given by $\mathscr{S}$ is actually a combinatorial geometry. The *zonotope* determined by $\mathscr{S}$ is the set of vectors

$$Z = \left\{ \mathbf{v} : \mathbf{v} = \sum_{i=1}^{q} \alpha_i \mathbf{x}_i, \quad \text{where } -1 \leqslant \alpha_i \leqslant 1 \quad \text{for all } i \right\}.$$

Figure 3.1. A zonotope, its vector star, and its matroid.

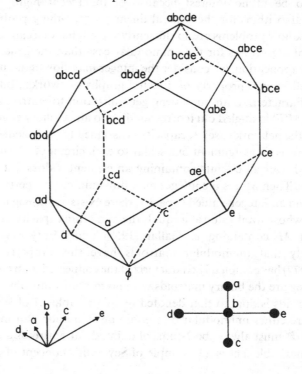

Equivalently, we may say that Z is the vector sum of the q line segments $L_i =$ convex hull $(-\mathbf{x}_i, \mathbf{x}_i)$. We call $\mathcal{S}$ the *vector star* of Z. Zonotopes are convex, centrally symmetric polytopes with many interesting properties. A three-dimensional example is given in Figure 3.1. In this example, *abc* and *bde* are chosen to be collinear. The vertices of the zonotope are vectors with each $\alpha_i = \pm 1$, and we have labelled each vertex by the vectors having $\alpha_i = +1$ at that vertex.

An interesting question is whether Z *packs* $\mathbb{E}^n$, (where n is the dimension of Z), that is, whether translates of Z may be placed to fill up $\mathbb{E}^n$ while intersecting each other only on their exterior faces. Shephard (1974) and McMullen (1975) have completely answered this question, *via* the following proposition. We assume that $\mathcal{S}$ spans $\mathbb{E}^n$.

3.3.4. Proposition. *A zonotope packs* $\mathbb{E}^n$ *if and only if its vector star is a binary matroid.*

But, in fact, the vector star is given as a vector matroid over the field $\mathbb{R}$. Hence by Theorem 3.1.1. condition (5), the vector star is binary if and only if it is unimodular.

The zonotope pictured in Figure 3.1 does satisfy the conditions of Proposition 3.3.4. so it does pack $\mathbb{E}^3$.

Exercises

3.1. Show that graphic and cographic matroids are signable.

3.2. Prove that a matroid $M(E)$ may be decomposed as a 2-sum of two matroids if and only if M has a 2-separation, that is, a partition (X_1, X_2) of E with $|X_1| \geqslant 2, |X_2| \geqslant 2, rX_1 + rX_2 \leqslant rE + 1$.

3.3. Show that the class of unimodular matroids is not *filtered* in the sense of Brylawski and Kelly (1980), that is, that there exist unimodular matroids of the same rank n which are not both submatroids of any unimodular matroid of rank n.

3.4. (Aigner 1979) If A is a locally unimodular coordinatization (over $\mathbb{Q}$) of a unimodular matroid $M(E)$, A is $n \times N$ where $n = \operatorname{rank} M, N = |E|$, then det $(AA)^t =$ the number of bases of M.

3.5. Prove that 1-, 2-, and 3-sums of unimodular matroids are unimodular. What are the corresponding operations on coordinatizing matrices?

3.6. Show that the binary coordinatization for R_{10} described in Exercise 1.7 is projectively equivalent to one in which each column has the same number of zeros. Thus determine that this matroid has a doubly transitive group of automorphisms. Use this information to show that R_{10} is unimodular, but neither graphic nor cographic.

3.7. If $M(E)$ is unimodular, $e \in E$, such that $M - e$ is isomorphic to R_{10}, show that e must be a loop, isthmus, or parallel element [i.e., M is the parallel extension of some element of $M - e$: see White (1986), p. 180)]. This is essentially all that is needed to check that R_{10} is a splitter for the class of unimodular matroids (see White 1986, Exercise 7.50).

3.8. Prove that for each vector $\mathbf{v}$ in the vector star of a zonotope Z, the set of edges of Z parallel to $\mathbf{v}$ form a 'zone', or minimal cut-set of the graph G determined by the edge-skeleton of Z, i.e., a bond in $M(G)$.

3.9. Prove that a zonotope in E^3 is space-filling if and only if all of its projections onto a plane orthogonal to a vector in its star yield tessalations (quadrilateral or hexagonal) of the plane.

3.10. Let $C_{k,n}$ denote the binary matroid determined by the binary matrix consisting of the $n \times n$ identity matrix next to an $n \times n$ matrix consisting of all cyclic shifts of a column of k ones followed by $n - k$ zeros. Show that $C_{2,n}$ is always graphic, that $C_{3,5}$ is R_{10}, and $C_{3,n}$ is not unimodular for all $n > 5$.

3.11. Is $C_{k,n}$ unimodular for any $k \geqslant 4, n > k$?

References

Aigner, M. (1979). *Combinatorial Theory*. Springer-Verlag, New York.

Bixby, R. (1979). On Reid's characterization of the matroids representable over $GF(3)$. *J. Comb. Theory Ser. B* **26**, 174–204.

Bixby, R.E. and Cunningham, W.H. (1980). Converting linear programs to network programs. *Math. of Operations Research*, **5**, 321–57.

Bixby, R.E. and Cunningham, W.H. (1981). Matroids, graphs. and 3-connectivity, in *Graph Theory and Related Topics* (J.A. Bondy and U.S.R. Murty, eds.), pp. 91–103. Academic Press, New York.

Brylawski, T. (1975). A note on Tutte's unimodular representation theorem. *Proc. Amer. Math. Soc.* **52**, 499–502.

Brylawski, T. and Kelly, D. (1980). *Matroids and Combinatorial Geometries*. Carolina Lecture Series, Dept. of Mathematics, University of North Carolina at Chapel Hill.

Crapo H. and Rota, G.-C. (1970). *Combinatorial Geometries*, preliminary edition. MIT Press, Cambridge, Mass.

Cunningham, W.H. and Edmonds, J. (1987). Decomposition of linear systems, in preparation.

Gallai, T. (1959). Uber reguläre Kettengruppen. *Acta Math. Acad. Sci. Hungar.* **10**, 227–40.

Heller, I. (1957). On linear systems with integral valued solutions. *Pacific J. Math.* **7**, 1351–64.

McMullen, P. (1975). Space-tiling zonotopes. *Mathematika* **22**, 202–11.

Minty, G.J. (1966). On the axiomatic foundations of the theories of directed linear graphs, electrical networks, and network-programming. *J. Math. and Mech.* **15**, 485–520.

Seymour, P.D. (1980). Decomposition of regular matroids. *J. Comb. Theory Ser. B* **28**, 305–60.

Seymour, P.D. (1977). The matroids with the max-flow-min-cut property. *J. Comb. Theory Ser. B* **23**, 189–222.

Shephard, G.C. (1974). Space-filling zonotopes. *Mathematika* **21**, 261–9.

Tutte, W.T. (1958). A homotopy theorem for matroids, I and II. *Trans. Amer. Math. Soc.* **88**, 144–74.

Tutte, W.T. (1960). An algorithm for determining whether a given binary matroid is graphic. *Proc. Amer. Math. Soc.* **11**, 905–17.

White, N., ed. (1986). *Theory of Matroids*. Cambridge University Press.

White, N., ed. (1988). *Combinatorial Geometries: Advanced Theory*. Cambridge University Press, to appear.

4

Introduction to Matching Theory

RICHARD A. BRUALDI

4.1. Matchings on Matroids

One of the many fascinating aspects of matching theory is the interdependence of the theorems of the subject. There are many theorems that can be regarded as basic, and it seems that one can begin the theory with any one of them and use it as an important step in the proofs of the others. It follows that a result derived as a consequence of another theorem should not necessarily be regarded only as a corollary. It may have just as much claim to centrality and importance as the theorem itself. Different approaches exhibit different relationship between the theorems. Thus we have considerable latitude in choosing our starting point and our progression through some of the important results of matching theory. We begin with the notion of a relation between two sets.

Let S_1 and S_2 be two finite sets and $R \subseteq S_1 \times S_2$ a relation between the elements of S_1 and those of S_2. We suppose, as we may without loss of generality, that S_1 and S_2 are disjoint. Thus the relation R can be modelled by means of a finite bipartite graph $\Gamma = \Gamma(S_1, S_2)$. The vertices of Γ are the elements of $S_1 \cup S_2$, and there is an edge $[x, y]$ joining $x \in S_1$ and $y \in S_2$ if and only if xRy. A *matching* (of cardinality k) of Γ is a set Θ of edges $[x_1, y_1], \ldots, [x_k, y_k]$ of Γ where $x_1, \ldots, x_k$ are distinct elements of S_1 and $y_1, \ldots, y_k$ are distinct elements of S_2. Thus the matching Θ is a set of pairwise vertex disjoint edges, and these edges *match* the subset $\{x_1, \ldots, x_k\}$ of S_1 with the subset $\{y_1, \ldots, y_k\}$ of S_2. A *separating set* of Γ is a set Z of vertices of Γ such that for each edge $[x, y]$ of Γ either $x \in Z$ or $y \in Z$. If $Z = (Z \cap S_1) \cup (Z \cap S_2)$ is a separating set of Γ, then there are no edges which join a vertex in $S_1 - (Z \cap S_1)$ and a vertex in $S_2 - (Z \cap S_2)$.

We now have the following fundamental result.

4.1.1. Proposition. (*D. König 1931*) *The maximum cardinality of a matching of* *the bipartite graph* $\Gamma(S_1, S_2)$ *equals the minimum cardinality of a separating set of* $\Gamma(S_1, S_2)$.

Let $\Gamma(S_1, S_2)$ be a bipartite graph and let $X \subseteq S_1$. We define ∂X to be the set of vertices y of S_2 such that $[x, y]$ is an edge of Γ for some $x \in X$. It is readily verified that for all $X \subseteq S_1$, $\partial X \cup (S_1 - X)$ is a separating set of Γ. A similar definition can be made for $Y \subseteq S_2$. Now let Z be a separating set of Γ with $Z_1 = Z \cap S_1$ and $Z_2 = Z \cap S_2$. Let $X = S_1 - Z_1$. Then it follows that $\partial X \subseteq Z_2$, and hence $\partial X \cup (S_1 - X)$ is a separating set contained in Z. Thus from Proposition 4.1.1 we obtain the following.

4.1.2. Proposition. *The maximum cardinality of a matching of the bipartite graph* $\Gamma(S_1, S_2)$ *equals*

$$\min\{|\partial X| + |S_1 - X| : X \subseteq S_1\}.$$

A number of interesting theorems can be deduced from Propositions 4.1.1 and 4.1.2. Let I and S be finite sets, and let $\mathscr{A}(I) = (A_i : i \in I)$ be a family of subsets of S indexed by I. A family $(x_i : i \in I)$ of elements of S is a *system of representatives* (SR) of $\mathscr{A}(I)$ if $x_i \in A_i (i \in I)$, and a *system of distinct representatives* (SDR) if in addition $x_i \neq x_j$ for $i, j \in I$ with $i \neq j$. A *transversal* of $\mathscr{A}(I)$ is a set T for which there exists a bijection $\sigma : T \to I$ such that $x \in A_{\sigma(x)}$ for $x \in T$. It follows that T is a transversal if and only if its elements can be indexed by I to form an SDR of $\mathscr{A}(I)$. A set P for which there exists an injection $\sigma : P \to I$ such that $x \in A_{\sigma(x)}$ for $x \in P$ is called a *partial transversal* of $\mathscr{A}(I)$. Thus a partial transversal of $\mathscr{A}(I)$ is a transversal of a subfamily $\mathscr{A}(J) = (A_i : i \in J)$ of $\mathscr{A}(I)$ for some $J \subseteq I$.

With the finite family $\mathscr{A}(I) = (A_i : i \in I)$ of subsets of the finite set S there is associated a finite bipartite graph $\Gamma_{\mathscr{A}} = \Gamma_{\mathscr{A}}(I, S)$ where we suppose, without loss of generality, that $I \cap S = \varnothing$. There is an edge joining $i \in I$ and $s \in S$ if and only if $s \in A_i$. It follows that for $J \subseteq I$, a set P is a partial transversal of the subfamily $\mathscr{A}(J)$ if and only if there is a matching of $\Gamma_{\mathscr{A}}$ which matches J with P. For $J \subseteq I$, $\partial J = \bigcup_{i \in J} A_i$ and we denote this set by $A(J)$.

From Proposition 4.1.2 we immediately obtain the following.

4.1.3. Proposition. *(O. Ore 1955) The maximum cardinality of a partial transversal of the finite family* $\mathscr{A}(I)$ *is given by*

$$min\{|A(J)| + |I - J| : J \subseteq I\}.$$

As a corollary we obtain the following result.

4.1.4. Proposition. *(P. Hall 1935) The finite family* $\mathscr{A}(I)$ *has a transversal if and only if*

$$|A(J)| \geqslant |J| \quad (J \subseteq I).$$

Propositions 4.1.3 and 4.1.4 can be generalized by assuming that $M(S)$ is a finite matroid and requiring the partial transversal or transversal to be an independent set of the matroid.

4.1.5. Proposition. (*H. Perfect 1969b*) *Let $M(S)$ be a finite matroid with rank function r, and let $\mathscr{A}(I)$ be a finite family of subsets of S. Then the maximum cardinality of a partial transversal of $\mathscr{A}(I)$ which is an independent set of $M(S)$ is given by*

$$min \{r(A(J)) + |I - J| : J \subseteq I\}.$$

A special case of this proposition is the following.

4.1.6. Proposition. (*R. Rado 1949*) *Let $M(S)$ be a finite matroid with rank function r, and let $\mathscr{A}(I) = (A_i : i \in I)$ be a finite family of subsets of S. Then $\mathscr{A}(I)$ has a transversal which is an independent set of $M(S)$ if and only if*

$$r(A(J)) \geqslant |J| \quad (J \subseteq I).$$

Note that Propositions 4.1.3 and 4.1.4 result from Propositions 4.1.5 and 4.1.6 respectively, when the matroid $M(S)$ is the free matroid on S whose rank function is the cardinality function. We also note that there is a kind of converse to Proposition 4.1.6 due to Rado (1949) which we discuss briefly. Let S be a non-empty set and let $\mathscr{I}$ be a non-empty collection of subsets of S. Suppose for each finite family $\mathscr{A}(I) = (A_i : i \in I)$ of subsets of S, one of the following conditions holds if and only if the other does:

(i) $\mathscr{A}(I)$ has a transversal which belongs to $\mathscr{I}$;
(ii) for each $J \subseteq I$, $A(J)$ contains a subset of cardinality $|J|$ which belongs to $\mathscr{I}$.

Then $\mathscr{I}$ is the collection of independent sets of a matroid $M(S)$. Its follows that in Proposition 4.1.6 the matroid structure is essential.

The above propositions can be further generalized and for this we return to the setting of bipartite graphs. Let $\Gamma(S_1, S_2)$ be a finite bipartite graph and let $M_1(S_1)$ and $M_2(S_2)$ be finite matroids with rank functions r_1 and r_2, respectively. We are now interested in matchings of Γ which match an independent subset of $M_1(S_1)$ with an independent subset of $M_2(S_2)$. Let Θ be a matching which matches the independent set X_1 of $M_1(S_1)$ with the independent set X_2 of $M_2(S_2)$, and let $Z = Z_1 \cup Z_2$ be a separating set of Γ where $Z_1 \subseteq S_1$ and $Z_2 \subseteq S_2$. Then for each edge $[x, y]$ of Θ either $x \in Z_1$ or $y \in Z_2$, and it follows that

$$|\Theta| \leqslant r_1(X_1 \cap Z_1) + r_2(X_2 \cap Z_2)$$
$$\leqslant r_1(Z_1) + r_2(Z_2).$$

4.1.7. Theorem. *The maximum cardinality m of a matching of the bipartite graph $\Gamma(S_1, S_2)$ which matches an independent set of $M_1(S_1)$ with an independent set of $M_2(S_2)$ equals n where*

$n = \min \{r_1(Z_1) + r_2(Z_2) : z_1 \subseteq S_1, Z_2 \subseteq S_2, Z_1 \cup Z_2 \text{ a separating set}\}$ *or, equivalently,*

$$n = \min \{r_1(S_1 - X) + r_2(\partial X) : X \subseteq S_1\}.$$

Proof. That the two expressions for n have the same value follows, since every separating set Z contains a separating set of the form $(S_1 - X) \cup \partial X$ for some $X \subseteq S_1$. The calculation above has shown that $m \leqslant n$, so that it suffices to show that there exists a matching of cardinality n of the required type. If $n = 0$ or 1, this is readily verified. Thus we suppose $n \geqslant 2$ and use induction on $|S_1 \cup S_2|$. It is convenient to consider two cases.

Case 1. The only separating sets $Z_1 \cup Z_2 (Z_1 \subseteq S_1, Z_2 \subseteq S_2)$ with $r_1(Z_1) + r_2(Z_2) = n$ satisfy $Z_1 = \varnothing$ or $Z_2 = \varnothing$.

Choose an edge $[x, y]$ of Γ such that $\{x\}$ is an independent set of $M_1(S_1)$ and $\{y\}$ is an independent set of $M_2(S_2)$. Let $T_1 = S_1 - \{x\}$ and $T_2 = S_2 - \{y\}$. Let $\Gamma^*(T_1, T_2)$ be the bipartite graph obtained from Γ by deleting the vertices x and y and all edges meeting x or y. Finally consider the contractions $M_1/\{x\}$ and $M_2/\{y\}$ with rank functions r_1^* and r_2^*, respectively. Let $Z_1^* \cup Z_2^*$ be a separating set of Γ^* where $Z_1^* \subseteq T_1$ and $Z_2^* \subseteq T_2$. Then $Z_1 \cup Z_2$ where $Z_1 = Z_1^* \cup \{x\}$ and $Z_2^* \cup \{y\}$ is a separating set of Γ with $Z_1 \neq \varnothing \neq Z_2$, and it follows that

$$n + 1 \leqslant r_1(Z_1) + r_2(Z_2) = r_1^*(Z_1^*) + 1 + r_2^*(Z_2^*) + 1$$

and thus

$$n - 1 \leqslant r_1^*(Z_1^*) + r_2^*(Z_2^*).$$

By the inductive assumption there exists a matching Θ^* of Γ^* of cardinality $n - 1$ which matches an independent set of $M_1/\{x\}$ with an independent set of $M_2/\{y\}$. It follows that $\Theta = \Theta^* \cup \{[x, y]\}$ is the required matching.

Case 2. There exists a separating set $Z_1 \cup Z_2 (Z_1 \subseteq S_1, Z_2 \subseteq S_2)$ with $r_1(Z_1) + r_2(Z_2) = n$ where $Z_1 \neq \varnothing \neq Z_2$.

First we consider the bipartite graph $\Gamma^*(Z_1, S_2 - Z_2)$ obtained from Γ by deleting the vertices of $S_1 - Z_1$ and those of Z_2 and all edges meeting at least one of these vertices. We also consider the matroids $M_1(Z_1)$ and M_2/Z_2 with rank functions r_1^* and r_2^*, respectively. Let $Z_1^* \cup Z_2^*$ be a separating set of Γ^* where $Z_1^* \subseteq Z_1$ and $Z_2^* \subseteq S_2 - Z_2$. Then $Z_1^* \cup (Z_2^* \cup Z_2)$ is a separating set of Γ, and it follows that

$$
\begin{aligned}
n &\leqslant r_1(Z_1^*) + r_2(Z_2^* \cup Z_2) \\
&\leqslant r_1^*(Z_1^*) + r_2^*(Z_2^*) + r_2(Z_2),
\end{aligned}
$$

and thus
$$r_1(Z_1) = n - r_2(Z_2) \leqslant r_1^*(Z_1^*) + r_2^*(Z_2^*).$$

It follows from the inductive assumption that there exists a matching Θ^* of Γ^* of cardinality $r_1(Z_1)$ which matches an independent set of $M_1(Z_1)$ with an independent set of M_1/Z_2. In a similar way we can define a graph Γ^{**} and obtain a matching Θ^{**} of Γ^{**} of cardinality $r_2(Z_2)$ which matches an independent set of M_1/Z_1 with an independent set of $M_2(Z_2)$. Then $\Theta = \Theta^* \cup \Theta^{**}$ is a matching of cardinality $n = r_1(Z_1) + r_2(Z_2)$ which matches an independent set of $M_1(S_1)$ with an independent set of $M_2(S_2)$.

Thus the theorem holds by induction. $\qquad\square$

Each of Propositions 4.1.1 to 4.1.6 is a special case of Theorem 4.1.7, so that all are now proved.

The notion of a matching of a bipartite graph can be extended to any finite graph $\Gamma = \Gamma(V)$. Here V is the finite vertex set of Γ, and the edges of Γ are unordered pairs of vertices $[x, y]$ with $x \neq y$. A *matching of* Γ is a set Θ of pairwise vertex disjoint edges. A set X of vertices is said to *meet the matching* Θ if each vertex in X is a vertex of a (unique) edge of Θ; there may be vertices not in X which are also vertices of edges of Θ. A *perfect matching* (also called a 1-*factor*) is a matching which V meets. Thus in a perfect matching Θ each vertex of Γ is a vertex of an edge of Θ. Clearly, a necessary condition for the graph Γ to have a perfect matching is that the number of its vertices be even. A less obvious necessary condition is that the deletion of k vertices of Γ and all edges meeting at least one of them results in a graph with at most k connected components with an odd number of vertices (there may be any number of connected components with an even number of vertices). In a remarkable discovery Tutte (1947) proved that this condition (for $k = 0, 1, \ldots, |V|$) is also sufficient for the existence of a perfect matching. We shall prove an extension of Tutte's theorem due to Berge (1958) and then deduce Tutte's theorem as a special case. Then we shall see how matching gives rise to matroids. To formulate this theorem we introduce the following notation.

Let $\Gamma = \Gamma(V)$ be a finite graph and let $S \subseteq V$. By $\Gamma(S)$ we denote the graph obtained from Γ by deleting all vertices not in S and all edges at least one of whose vertices is not in S. Note that if $T \subseteq S$ and $\Gamma_1 = \Gamma(S)$, then $\Gamma(T) = \Gamma_1(T)$. For $S \subseteq V$, the graph $\Gamma(V - S)$ has in general several connected components. We denote by $p(\Gamma; S)$ the number of *odd components* of $\Gamma(V - S)$, that is, the number of connected components of $\Gamma(V - S)$ having an odd number of vertices. We note that for all $S \subseteq V, |V| + |S| - p(\Gamma; S)$ is even.

4.1.8. Proposition. *Let $\Gamma(V)$ be a finite graph. Then the maximum cardinality α of a matching satisfies*

$$2\alpha = \min\{|V| + |S| - p(\Gamma; S) : S \subseteq V\}.$$

Proof. Denote the above minimum by $2m$. Let Θ be a matching and let $S \subseteq V$. Each odd component of $\Gamma(V - S)$ contains a vertex x such that either x does not meet Θ or else there exists a vertex $s \in S$ such that $[x, s]$ is an edge of Θ. Hence the number of vertices which meet Θ is at most

$$|V| - (p(\Gamma; S) - |S|) = 2m$$

and it follows that

$$2|\Theta| \leqslant 2m.$$

Hence $\alpha \leqslant m$. We now prove by induction on m that Γ has a matching of cardinality m, from which the proposition follows. The case $m = 0$ being obvious, we suppose $m \geqslant 1$.

Let τ be the collection of maximal subsets T of V for which

$$|V| + |T| - p(\Gamma; T) = 2m.$$

Let $T \in \tau$ and suppose that $\Gamma(V - T)$ has an even component (a component with an even number of vertices). Let x be any vertex of such a component. Then $p(\Gamma; T \cup \{x\}) \geqslant p(\Gamma; T) + 1$ and hence

$$|V| + |T \cup \{x\}| - p(\Gamma; T \cup \{x\}) \leqslant |V| + |T| - p(\Gamma; T).$$

It follows that equality holds above and we contradict $T \in \tau$. Thus for all $T \in \tau$, $\Gamma(V - T)$ has no even components. We now distinguish two cases.

Case 1. There exists $T \in \tau$ such that either $T \neq \emptyset$ or $\Gamma(V - T)$ has at least two odd components with more than one vertex.

Let $\Gamma_i = \Gamma(T_i) \ (i \in I)$ be the odd components of $\Gamma(V - T)$. Thus $|I| = p(\Gamma; T)$, the $T_i \ (i \in I)$ are disjoint sets with an odd number of vertices, and

$$2m = \sum_{i \in I} (|T_i| - 1) + 2|T|.$$

Because of our assumption in this case, $|T_i| - 1 \leqslant 2(m - 1) \ (i \in I)$. Let i be an arbitrary but fixed element of I, and suppose that for some $z \in T_i$, the graph $\Gamma_i' = \Gamma_i(T_i - \{z\})$ did not have a matching of cardinality $(|T_i| - 1)/2$. By the inductive assumption there exists $S \subseteq T_i - \{z\}$ such that

$$|T_i| - 1 > |T_i| - 1 + |S| - p(\Gamma_i'; S)$$

or, equivalently,

$$p(\Gamma_i'; S) > |S|.$$

Since Γ_i' has an even number of vertices, it follows that

$$p(\Gamma_i'; S) \geqslant |S| + 2.$$

We then calculate that

$$\begin{aligned} p(\Gamma; T \cup S \cup \{z\}) &= p(\Gamma; T) - 1 + p(\Gamma_i'; S) \\ &\geqslant |V| + |T| - 2m - 1 + |S| + 2 \\ &\geqslant |V| + |T \cup S \cup \{z\}| - 2m. \end{aligned}$$

It follows that equality holds throughout, and we contradict the fact that $T \in \tau$. Hence for each $i \in I$ and each $z \in T_i$, Γ'_i has a perfect matching $\Theta_i(z)$.

We now consider the bipartite graph $\Gamma^* = \Gamma^*(T, I)$ whose edges are the pairs $[t, i]$ such that $t \in T$, $i \in I$, and there is an edge of Γ joining t and some vertex in T_i. Suppose that Γ^* did not have a matching which T meets (that is, a matching of cardinality equal to $|T|$). It then follows from Proposition 4.1.6 (or Proposition 4.1.2) that there exists a set $S \subseteq T$ and a set $J \subseteq I$ such that $|J| < |S|$ and no edge of Γ^* is of the form $[s, i]$ where $s \in S$ and $i \in I - J$. We then conclude that

$$p(\Gamma; T - S) \geqslant |I| - |J| = p(\Gamma; T) - |J|,$$

and we calculate that

$$|V| + |T - S| - p(\Gamma; T - S) \leqslant |V| + |T| - |S| - p(\Gamma; T) + |J|$$
$$\leqslant 2m - (|S| - |J|)$$
$$< 2m.$$

From this contradiction we conclude that Γ^* has a matching which T meets. Thus there exist $K \subseteq I$ and $z_i \in T_i$ for $i \in K$ such that Γ has a matching Θ which matches T with $\{z_i : i \in K\}$. For $i \in I - K$, let z_i be any element of T_i. Then

$$\Theta \cup \left(\bigcup_{i \in I} \Theta_i(z_i) \right)$$

is a matching of Γ having cardinality

$$|T| + \sum_{i \in I} (|T_i| - 1)/2 = m$$

Case 2. $\tau = \{\varnothing\}$ and Γ has exactly one odd component with more than one vertex.

Thus for all $S \subseteq V$ with $S \neq \varnothing$,

$$2m < |V| + |S| - p(\Gamma; S),$$

and hence

$$2m + 2 \leqslant |V| + |S| - p(\Gamma; S),$$

while $2m = |V| - p(\Gamma; \varnothing)$, where $p(\Gamma; \varnothing)$ is the number of (odd) components of Γ. The component of Γ with more than one vertex has $2m + 1$ vertices. Let x and y be any pair of vertices such that $[x, y]$ is an edge, and consider the graph $\Gamma^* = \Gamma(V^*)$ where $V^* = V - \{x, y\}$. Let $S^* \subseteq V^*$ and let $S = S^* \cup \{x, y\}$. Then $p(\Gamma; S) = p(\Gamma^*; S^*)$ and, since $S \neq \varnothing$,

$$2m + 2 \leqslant |V| + |S| - p(\Gamma; S)$$
$$\leqslant |V^*| + 2 + |S^*| + 2 - p(\Gamma^*, S^*)$$

so that

$$2(m - 1) \leqslant |V^*| + |S^*| - p(\Gamma^*, S^*).$$

Since this inequality holds for all $S^* \subseteq V^*$, it follows from the inductive hypothesis that Γ^* has a matching Θ^* of cardinality $m - 1$. Hence $\Theta^* \cup \{[x, y]\}$ is a matching of G of cardinality m.

The proof of the Proposition is now complete. $\square$

In proving Proposition 4.1.8 we have proved more than the statement that the maximum number m of edges in a matching of a finite graph Γ equals

$$\tfrac{1}{2} \min \{|V| + |S| - p(\Gamma; S) : S \subseteq V\}.$$

We have, in addition, described the structure of the matchings of Γ of cardinality m.

4.1.9. Proposition. *Let $\Gamma(V)$ be a finite graph, and let m be the maximum cardinality of a matching of Γ. Then there exists $T \subseteq V$ such that the connected components of Γ $(V - T)$ have vertex sets $T_i (i \in I)$ of odd cardinality and the following property holds.*

Let $\Gamma^*(T, I)$ be the bipartite graph such that for $t \in T$ and $i \in I$, $[t, i]$ is an edge if and only if $[t, x]$ is an edge of Γ for some $x \in T_i$. Then every matching of Γ of cardinality m is obtained in the following way:

(i) *Choose a matching $\Theta^* = \{[t, i_t] : t \in T\}$ of cardinality $|T|$ of Γ^*.*

(ii) *For each $j \in I$ choose a vertex $z_j \in T_j$ such that when $j = i_t$ for some $t \in T$, $[t, z_{i_t}]$ is an edge of Γ. Let $\Theta = \{[t, z_{i_t}] : t \in T\}$.*

(iii) *For each $i \in I$ choose a matching Θ_i of $\Gamma(T_i)$ of cardinality $(|T_i| - 1)/2$ such that z_i does not meet Θ_i.*

(iv) *Then $\Theta \cup (\bigcup_{i \in I} \Theta_i)$ is a matching of cardinality m.*

Moreover, for any choice of Θ^* satisfying (i) and any choice of Θ satisfying (ii), there exists $\Theta_i (i \in I)$ satisfying (iii).

Proof. Suppose T satisfies Case 1 of the proof of Proposition 4.1.8. Then in examining the proof of Proposition 4.1.8 we see that steps (i), (ii), and (iii) can be carried out and in carrying them out we always obtain a matching of cardinality m as described in (iv). Moreover, it follows from considerations of cardinality that every matching of cardinality m arises by carrying out steps (i) to (iv). Now suppose that T satisfies Case 2 of the proof of Proposition 4.1.8. Then, in particular, $T = \varnothing$. Then steps (i) and (ii) above are vacuous, and exactly one of the sets $T_i (i \in I)$ has more than one vertex. If T_k is this set, then a proof like that used in Case 1 shows that for each $z_k \in T_k$ there is a matching of $\Gamma(T_k)$ of cardinality $(|T_k| - 1)/2$ which z_k does not meet. The Proposition now follows. $\square$

As a special case of Proposition 4.1.8 we obtain the following.

4.1.10. Proposition. *(W.T. Tutte 1947) Let $\Gamma(V)$ be a finite graph. Then Γ has a* *perfect matching if and only if*

$$p(\Gamma; S) \leqslant |S| \quad (S \subseteq V)$$

Proof. By taking $S = \varnothing$ in the inequality, we see that Γ has an even number of vertices. The theorem now readily follows from Proposition 4.1.8.   □

Let $\Gamma(V)$ be a finite graph and let $S, A \subseteq V$. We define $p(\Gamma; S, A)$ to be the number of odd components of $\Gamma(V - T)$ the set of whose vertices is a subset of A. Thus $p(\Gamma; S) = p(\Gamma; S, V)$. We now obtain the following generalization of Proposition 4.1.8.

4.1.11. Proposition. (*R.A. Brualdi 1971*) *Let $\Gamma(V)$ be a finite graph and let $A \subseteq V$. Then the maximum cardinality of a subset of vertices of A which meets a matching of Γ equals*

$$min \{|A| + |S| - p(\Gamma; S, A) : S \subseteq V\}.$$

Proof. We first show that A meets a matching of Γ if and only if

$$p(\Gamma; S, A) \leqslant |S| \quad (S \subseteq V).$$

That this inequality must hold if A meets a matching of Γ is readily verified, and we turn to the converse. Let the maximum cardinality of a matching of Γ be m. Let W be a set with $V \cap W = \varnothing$ and $|W| = |V| - 2m$. Let $\Gamma^*(V \cup W)$ be the graph obtained from Γ by including as vertices the elements of W and including as edges the pairs $[x, y]$ whenever $x \in W$, $y \in V - A$ or $x, y \in W$ with $x \neq y$. Then Γ^* has an even number of vertices, and it is straightforward to check that A meets a matching of Γ if and only if Γ^* has a perfect matching.

Suppose $p(\Gamma; S, A) \leqslant |S| (S \subseteq V)$. Let $T \subseteq V \cup W$. If $W \subseteq T$, then by Proposition 4.1.8,

$$p(\Gamma^*; T) = p(\Gamma; T \cap V) \leqslant |V| + |T \cap V| - 2m$$
$$\leqslant |W| + |T \cap V|$$
$$\leqslant |T|.$$

Now suppose $W \nsubseteq T$. Then it follows from the definition of Γ^* that

$$p(\Gamma^*; T) \leqslant p(\Gamma; T \cap V; A) + 1$$
$$\leqslant |T \cap V| + 1.$$

Since Γ^* has an even number of vertices,

$$p(\Gamma^*; T) \leqslant |T|.$$

Hence, by Proposition 4.1.10, Γ^* has a perfect matching so that Γ has a matching which A meets.

Now let t be a non-negative integer with $t \leqslant |A|$. To prove the proposition it suffices to show that there exists a matching of Γ which at least t vertices of A meet if and only if

$$|A| + |S| - p(\Gamma; S, A) \geqslant t \quad (S \subseteq V).$$

It is readily verified that this inequality holds if t vertices of A meet a matching of Γ. To prove the converse, we suppose the inequality holds. If $t = |A|$, then the converse has been proved above. So suppose that $t < |A|$. We first consider the case where there is no edge $[x, y]$ in Γ where $x \in A$ and $y \in V - A$. For $S \subseteq A$,

$$p(\Gamma(A); S) = p(\Gamma; S, A) \leqslant |A| + |S| - t.$$

It now follows from Proposition 4.1.8 that $\Gamma(A)$ has a matching of cardinality at least $t/2$ and hence there exists a matching of Γ which at least t vertices of A meet. We now assume that there is at least one edge of Γ which joins a vertex of A to a vertex not in A. Let W be a set with $V \cap W = \varnothing$ and $|W| = |A| - t$. Let $\Gamma^*(V \cup W)$ be the graph obtained from Γ by including as vertices the elements of W and by including as edges the pairs $[x, y]$ whenever $x \in W$ and $y \in A$. It is readily verified that there is a matching of Γ which at least t vertices of A meet if and only if $A \cup W$ meets a matching of Γ^*. From the first part of the proof it suffices to show that

$$p(\Gamma^*; T, A \cup W) \leqslant |T| \quad (T \subseteq V \cup W).$$

Let $T \subseteq V \cup W$. If $A \subseteq T$, then

$$p(\Gamma^*; T, A \cup W) = |W - T| \leqslant |W| = |A| - t \leqslant |A| \leqslant |T|.$$

If $W \subseteq T$ then

$$p(\Gamma^*; T, A \cup W) = p(\Gamma; T - W, A) \leqslant |A| + |T - W| - t = |T|.$$

Thus we may suppose that $A \nsubseteq T$ and $W \nsubseteq T$. From the definition of Γ^* it follows that $p(\Gamma; T, A) = 0$ or 1. Hence if $T \neq \varnothing$,

$$p(\Gamma^*; T, A \cup W) \leqslant 1 \leqslant |T|.$$

If $T = \varnothing$, then since it is assumed that there is an edge joining a vertex in A to a vertex not in A,

$$p(\Gamma^*; T, A \cup W) = 0 = |T|.$$

Thus $A \cup W$ meets a matching of Γ^* and hence there is matching of Γ which at least t vertices of A meet. $\square$

4.2. Matching Matroids

We now show how matchings can be used to construct some interesting matroids whose rank functions can be obtained from previous theorems. Let

$\Gamma(V)$ be a finite graph and denote by $\mathscr{I}_\Gamma$ the collection of subsets of V which meet some matching of Γ. We then have the following.

4.2.1. Theorem. (*Edmonds and Fulkerson 1965*). *For any finite graph* $\Gamma(V)$, $\mathscr{I}_\Gamma$ *is the collection of independent sets of a matroid* $M_\Gamma(V)$.

Proof. We prove the theorem by showing that $\mathscr{I}_\Gamma$ satisfies the independence axioms for a matroid. Since (I_1) and (I_2) are obvious, we direct our attention to verifying (I_3). Let U_1 and U_2 be in $\mathscr{I}_\Gamma$ with $|U_1| < |U_2|$. There exist matchings Θ_1 and Θ_2 of Γ such that U_1 meets Θ_1 and U_2 meets Θ_2. If some vertex $x \in U_2 - U_1$ meets Θ_1, then $U_1 \cup \{x\}$ meets Θ_1 and hence $U_1 \cup \{x\} \in \mathscr{I}_\Gamma$. Thus we may suppose that no vertex $x \in U_2 - U_1$ meets Θ_1. We consider the subgraph Γ^* of Γ whose edges are the edges of $E = \Theta_1 \cup \Theta_2$ and whose vertices are the vertices of the edges in E. Then each vertex of Γ^* meets either one or two edges of Γ^*, and it follows that the connected components of Γ^* are either elementary chains joining two distinct vertices or elementary cycles of even length; in either case, since Θ_1 and Θ_2 are matchings, the edges alternate between Θ_1 and Θ_2. Since no $x \in U_2 - U_1$ meets Θ_1, it follows that a connected component of Γ^* which is a cycle, contains at least as many vertices of U_1 as of U_2. The components of Γ^* which are chains are of one of three types, determined by the nature of the first and last edges: (i) both are edges of Θ_1, (ii) both are edges of Θ_2, (iii) the first is an edge of Θ_1 and the last is an edge of Θ_2 (or vice versa). Type (i) chains contain at least as many vertices of U_1 as of U_2. The vertices of U_1 that belong to a type (ii) chain γ are vertices of the edges of Θ_2 that belong to γ. If the first vertex of a type (iii) chain γ belongs to U_1, then γ contains at least as many vertices of U_1 as of U_2; otherwise all vertices of γ that belong to U_1 are vertices of the edges of Θ_2 that belong to γ. Since $|U_1| < |U_2|$, it follows that there is a chain γ^* which is a component of Γ^* having the properties that it contains more vertices of U_2 than of U_1 and each vertex of U_1 which belongs to γ^* is a vertex of an edge of Θ_2. Let Θ_1' be the edges of γ^* which belong to Θ_1, and let Θ_2' be the edges of γ^* which belong to Θ_2. Then $\Theta_3 = (\Theta_1 - \Theta_1') \cup \Theta_2'$ is a matching for which there is a vertex $x \in U_2 - U_1$ such that $U_1 \cup \{x\}$ meets Θ_3. Thus $U_1 \cup \{x\} \in \mathscr{I}_\Gamma$, (I_3) holds, and the theorem is proved. $\square$

Proposition 4.1.11 furnishes an explicit formula for the rank function of $M_\Gamma(V)$, and we state this fact as a corollary.

4.2.2. Corollary. *Let* $\Gamma(V)$ *be a finite graph, and let* r *denote the rank function of the matroid* $M_\Gamma(V)$. *Then for* $A \subseteq V$,

$$r(A) = \min \{|A| + |S| - p(\Gamma; S, A) : S \subseteq V\}.$$

We call the matroid $M_\Gamma(V)$ *the matching matroid* of the finite graph

$\Gamma(V)$, and define *a matching matroid* to be any restriction $M_\Gamma(S)$ of the matching matroid of a finite graph. Thus in a matching matroid $M_\Gamma(S)$, S is a subset of the vertices of a graph. There may be edges of the graph having one of their two vertices outside of S [edges having neither vertex in S can be deleted from the graph with no change in $M_\Gamma(S)$]. A special (see, however, Proposition 4.2.6) kind of matching matroid is a transversal matroid defined as follows. Let $\mathscr{A}(I) = (A_i : i \in I)$ be a finite family of subsets of a finite set S, indexed by I, and let $\Gamma_\mathscr{A}(I, S)$ be the associated bipartite graph. Then the matching matroid of $\Gamma_\mathscr{A}$ restricted to S, $M_{\Gamma_\mathscr{A}}(S)$, is a matroid whose independent sets are the partial transversals of $\mathscr{A}$. We call such a matroid the *transversal matroid* of the family $\mathscr{A}$ of subsets of S and denote it by $M_\mathscr{A}(S)$. (If one restricts the matching matroid of $\Gamma_\mathscr{A}$ to I, one obtains a matroid whose independent sets are the subsets J of I such that the subfamily $\mathscr{A}(J)$ has a transversal. Such a matroid is the transversal matroid of the family $\mathscr{B}(S) = (B_s : s \in S)$ of subsets of I where $B_s = \{i : i \in I, s \in A_i\}$ for $s \in S$.) Note that for $X \subseteq S$, the restriction $M_{\Gamma_\mathscr{A}}(X)$ is a transversal matroid, indeed it is the transversal matroid of the family $\mathscr{A}(I) = (A_i \cap X : i \in I)$ of subsets of X. From Proposition 4.1.3 we can obtain a formula for the rank function of a transversal matroid.

4.2.3. Proposition. *Let $\mathscr{A}(I) = (A_i : i \in I)$ be a finite family of subsets of a finite set S, and let r denote the rank function of the transversal matroid $M_\mathscr{A}(S)$. Then for $X \subseteq S$,*

$$r(X) = \min\{|A(J) \cap X| + |I - J| : J \subseteq I\}.$$

Proof. Let $X \subseteq S$. Then $r(X)$ is the maximum cardinality of a subset of X which is a partial transversal of $\mathscr{A}$ and this equals the maximum cardinality of a partial transversal of the family $(A_i \cap X : i \in I)$. The formula now follows from Proposition 4.1.3. $\square$

Matching matroids arise from matchings in graphs, while transversal matroids arise from matchings in bipartite graphs. Since bipartite graphs are, in general, more elementary combinatorial objects than graphs, one might expect that transversal matroids would constitute a small subclass of the class of matching matroids. As a matter of fact these two classes of matroids are identical. Before deriving this result of Edmonds and Fulkerson from the structure of maximum cardinality matchings as given in Proposition 4.1.9, we prove two lemmas.

4.2.4. Lemma. *Let $M(S)$ be a finite matroid, and let Z be a collection of isthmuses of $M(S)$. Then $M(S)$ is a transversal matroid if and only if the restriction $M(S - Z)$ is a transversal matroid.*

Proof. We have already noted that any restriction of a transversal matroid is a transversal matroid. Suppose $M(S - Z)$ is the transversal matroid of the family $(A_i : i \in I)$ of subsets of $S - Z$. Let K be a set with $K \cap I = \varnothing$ and $|K| = |Z|$, and consider the family $\mathscr{B}(K \cup I)$ of subsets of S where $B_j = Z$ if $j \in K$ and $B_j = A_j$ if $j \in I$. Since each basis of $M(S)$ is the union of Z and a basis of $M(S - Z)$, it readily follows that $M(S)$ is the transversal matroid of $\mathscr{B}(K \cup I)$. $\qquad\square$

Let $M(S)$ be a finite matroid and let $x \in S$. Let E be a set such that $E \cap S = \varnothing$. Recall that the *series extension* of $M(S)$ at x by E is the matroid $M_1(S \cup E)$ whose bases are the following sets:
  (i) $B \cup E$, where B is a basis of $M(S)$;
  (ii) $B \cup \{x\} \cup (E - \{y\})$, where B is a basis of $M(S)$ not containing x and $y \in E$.
It follows that each basis of $M_1(S \cup E)$ contains all but at most one element of $E \cup \{x\}$. Moreover, for each $y \in E$, $M_1(S \cup E)$ is the series extension of $M_1(S - x + y)$ at y by $E - y + x$, where $S - x + y$ denotes $(S - \{x\}) \cup \{y\}$. A sequence of series extensions at distinct elements of E can be done in any order.

4.2.5. Lemma. *The series extension $M_1(S \cup E)$ is a transversal matroid if and only if $M(S)$ is a transversal matroid.*

Proof. Since $M(S) = M_1(S)$, it follows that $M(S)$ is a transversal matroid if $M_1(S \cup E)$ is. Suppose $M(S)$ is the transversal matroid of the family $\mathscr{A}(I) = (A_i : i \in I)$ of subsets of S. Let K be a set such that $K \cap I = \varnothing$ and $|K| = |E|$. Consider the family $\mathscr{B}(K \cup I)$ of subsets of $S \cup E$ where $B_j = E \cup \{x\}$ if $j \in K$, $B_j = A_j \cup E$ if $j \in I$ and $x \in A_j$, and $B_j = A_j$ if $j \in I$ and $x \notin A_j$. It is straightforward to check that $M_1(S \cup E)$ is the transversal matroid of $\mathscr{B}(K \cup I)$. $\qquad\square$

4.2.6. Proposition. (*Edmonds and Fulkerson 1965*) *A matching matroid is a transversal matroid.*

Proof. Since a matching matroid is a restriction of the matching matroid of a graph and since a restriction of a transversal matroid is a transversal matroid, it suffices to show that the matching matroid of a graph is a transversal matroid. Let $\Gamma(V)$ be a finite graph. Let $\Gamma^*(T, I)$ be the bipartite graph described in Proposition 4.1.9, whose notation we freely use. For each $i \in I$, choose $x_i \in T_i$, and let $S = \{x_i : i \in I\}$. Let $\mathscr{A}(T) = (A_t : t \in T)$ be the family of subsets of S where for $t \in T$, $A_t = \{x_i : [t, i] \text{ is an edge of } \Gamma^*\}$. Then it follows from Proposition 4.1.9 that the matroid $M_\Gamma(V - T)$ is obtained from the transversal matroid $M_{\mathscr{A}}(S)$ by the sequence of series extensions at x_i by $T_i - \{x_i\}$ ($i \in I$). It now follows from Lemma 4.2.5 that $M_\Gamma(V - T)$ is a transversal matroid. From Proposition 4.1.9 we see that T is a collection of isthmuses of $M_\Gamma(V)$. Hence by Lemma 4.2.4 the matching matroid $M_\Gamma(V)$ is a transversal matroid. $\qquad\square$

The construction for transversal matroids can be generalized to show how a matroid may induce a new matroid across a relation or bipartite graph. Let $\Gamma(S_1, S)$ be a finite bipartite graph and let $M(S)$ be a finite matroid whose rank function is denoted by r. Let Θ be a matching of Γ which matches the subset A_1 of S_1 with the subset A of S. We call Θ an *M-matching* provided A is an independent set of $M(S)$. In case $M(S)$ is a free matroid, every matching of Γ is an M-matching.

4.2.7. Theorem. *The collection $\mathscr{I}_1$ of subsets of S_1 which meet an M-matching of $\Gamma(S_1, S)$ are the independent sets of a matroid $M_1(S_1)$ with rank function r_1 where for each $X \subseteq S_1$,*

$$r_1(X) = min\,\{r(\partial A) + |X - A| : A \subseteq X\}.$$

Proof. We verify that $\mathscr{I}_1$ satisfies the independence axiom (I_3), axioms (I_1) and (I_2) being obvious. Let $U_1, T_1 \in \mathscr{I}_1$ where $|U_1| < |T_1|$. Thus there exist matchings Θ and Ω where Θ matches U_1 with an independent set U of $M(S)$ and Ω matches T_1 with an independent set T of $M(S)$. We may suppose that Θ and Ω have been chosen so that $|U \cap T|$ is as large as possible. The bipartite graph Γ_1 whose edges are those edges of Γ which belong to Θ or Ω and whose vertices are the vertices of these edges has connected components which are either elementary chains joining distinct vertices or elementary cycles of even length. Since $|U_1| < |T_1|$, $|U| < |T|$ and it follows from axiom (I_3) for $M(S)$ that there exists $t \in T - U$ such that $U \cup \{t\}$ is an independent set of $M(S)$. This t meets an edge of Ω but not an edge of Θ, and it follows that there is a connected component of Γ_1 which is an elementary chain γ joining t to some vertex x where $x \in U - T$ or $x \in T_1 - U_1$. Suppose that $x \in U - T$. Let $\Theta' = (\Theta - \gamma_\Theta) \cup \gamma_\Omega$ where γ_Θ consists of those edges of γ which belong to Θ and γ_Ω consists of those that belong to Ω. Then Θ' is a matching of Γ which matches U_1 with $U - x + t$. Since $U \cup \{t\}$ is an independent set of $M(S)$, so is $U - x + t$. Hence Θ' is an M-matching. Since $|(U - x + t) \cap T| = 1 + |U \cap T|$, we have a contradiction. Therefore $x \in T_1 - U_1$. The matching Θ' defined above then matches $U \cup \{x\}$ with $U \cup \{t\}$. It follows that Θ' is an M-matching and hence $U_1 \cup \{x\} \in \mathscr{I}_1$. Hence axiom (I_3) is satisfied, and $\mathscr{I}_1$ is the collection of independent sets of a matroid $M_1(S_1)$. The formula for the rank function of $M_1(S_1)$ follows readily from Proposition 4.1.2 (or Proposition 4.1.5). $\square$

4.3. Applications

To conclude we give several applications of some of the previous theorems. Let $\mathscr{A}(I) = (A_i : i \in I)$ be a finite family of subsets of a finite set S. Then the collection of partial transversals of $\mathscr{A}$ are the independent sets of a matroid $M_{\mathscr{A}}(S)$. Suppose $\mathscr{A}$ has a transversal so that the bases of $M_{\mathscr{A}}(S)$ are the transversals of $\mathscr{A}$. Since every independent set of a matroid can be enlarged to a basis, we obtain the following result first proved by Hoffman and Kuhn (1956) before the discovery of transversal matroids.

4.3.1. Proposition. *Let* $\mathscr{A}(I) = (A_i : i \in I)$ *be a finite family of subsets of a finite set* S, *and let* $X \subseteq S$. *Then* $\mathscr{A}$ *has a transversal containing* X *if and only if* $\mathscr{A}$ *has a transversal and* X *is a partial transversal of* $\mathscr{A}$.

Let r denote the rank function of $M_{\mathscr{A}}(S)$. Then X is a partial transversal of $\mathscr{A}$ if and only if $r(X) = |X|$. Thus Propositions 4.1.4 and 4.2.3 in combination Proposition 4.3.1 furnish criteria for X to be a subset of a transversal of $\mathscr{A}$.

4.3.2. Proposition. (*Ford and Fulkerson 1958*) *Let* $\mathscr{A}(I) = (A_i : i \in I)$ *and* $\mathscr{B}(I) = (B_i : i \in I)$ *be finite families of subsets of a finite set* S. *Then there exists* $T \subseteq S$ *such that* T *is a transversal of both* $\mathscr{A}(I)$ *and* $\mathscr{B}(I)$ *if and only if*

$$|A(J) \cap B(K)| \geqslant |J| + |K| - |I| \quad (J, K \subseteq I).$$

Proof. we consider the transversal matroid $M_{\mathscr{A}}(S)$ whose independent sets are the partial transversals of $\mathscr{A}$. Then $\mathscr{A}$ and $\mathscr{B}$ have a common transversal T if and only if $\mathscr{B}$ has a transversal which is an independent set of $M_{\mathscr{A}}(S)$. Let r denote the rank function of $M_{\mathscr{A}}(S)$. Then by Proposition 4.1.6 there is an independent set of $M_{\mathscr{A}}(S)$ which is a transversal of $\mathscr{B}$ if and only if

$$r(B(K)) \geqslant |K| \quad (K \subseteq I).$$

It follows from Proposition 4.2.3 by setting $X = B(K)$ that the previous inequality holds for all $K \subseteq I$ if and only if

$$|A(J) \cap B(K)| + |I - J| \geqslant |K| \quad (J, K \subseteq I),$$

and the proposition follows. □

The previous proof can be easily modified to obtain a criterion for there to exist a set T of prescribed cardinality m which is a common partial transversal of $\mathscr{A}(I_1)$ and $\mathscr{B}(I_2)$, both finite families of subsets of S, equivalently for there to exist a common independent set of cardinality m of the two matroids $M_{\mathscr{A}}(S)$ and $M_{\mathscr{B}}(S)$. More generally we have the following result attributed to Edmonds (1970).

4.3.3. Proposition. *Let* $M_1(S)$ *and* $M_2(S)$ *be two finite matroids with rank functions* r_1 *and* r_2, *respectively. Let* m *be a non-negative integer. Then there exists* $T \subseteq S$ *such that* T *is an independent set of both matroids* $M_1(S)$ *and* $M_2(S)$ *if and only if*

$$r_1(X) + r_2(S - X) \geqslant m \quad \text{for all} \quad X \subseteq S.$$

Proof. Briefly, we consider the bipartite graph Γ whose vertices consist of two disjoint 'copies' of S where the only edges are the $|S|$ edges joining the two copies of each element of S. The result now follows easily from Theorem 4.1.7. $\square$

Finally, we indicate a proof of the following.

4.3.4. Proposition. (*Edmonds and Fulkerson 1965*) *Let* $M_1(S), \ldots, M_k(S)$ *be finite matroids with rank functions* $r_1, \ldots, r_k$, *respectively. Then there exist pairwise disjoint subsets* $B_1, \ldots, B_k$ *of* S *such that* B_i *is a basis of* $M(S_i)$ *for* $i = 1, \ldots, k$ *if and only if*

$$\sum_{i=1}^{k} r_i(A) + |S - A| \geqslant \sum_{i=1}^{k} r_i(S) \quad (A \subseteq S).$$

Proof. Let $S_1, \ldots, S_k$ be 'copies' of S such that $S, S_1, \ldots, S_k$ are pairwise disjoint. Consider the bipartite graph $\Gamma(S, S_1 \cup \cdots \cup S_k)$ where for each $s \in S$ there is an edge joining s to each of its k 'copies'. Let $M(S_1 \cup \cdots \cup S_k)$ be the matroid which is the direct sum of the matroids $M_1(S_1), \ldots, M_k(S_k)$. The rank function r of this matroid satisfies

$$r(A_1 \cup \cdots \cup A_k) = r_1(A_1) + \cdots + r_k(A_k) \quad (A_i \subseteq S_i, i = 1, \ldots, k).$$

It follows easily that there exist pairwise disjoint bases if $M_1(S), \ldots,$ $M_k(S)$ if and only if Γ has a matching which matches a subset of S of cardinality $r_1(S) + \cdots + r_k(S)$ with an independent set of $M(S_1 \cup \cdots \cup S_k)$. Using Theorem 4.1.7, we see that such a matching exists if and only if

$$\sum_{i=1}^{k} r_i(A^{(i)}) + |S - A| \geqslant \sum_{i=1}^{k} r_i(S) \quad (A \subseteq S),$$

where $A^{(i)}$ is the 'copy' of A in S_i. Since $r_i(A^{(i)}) = r_i(A) \; (i = 1, \ldots, k)$, the result now follows. $\square$

Notes

Theorem 4.1.7 appeared in an unpublished manuscript (Symmetrized form of R. Rado's theorem on independent representatives, 1967) by Brualdi. It occurs in a more general form in Brualdi (1970) and also in Aigner and Dowling (1971). The proof given here follows that in the unpublished manuscript. There are many known proofs of Rado's theorem (Proposition 4.1.6) and Hall's theorem (Proposition 4.1.4). The reader is referred to Mirsky (1971) and the many references therein and also to Welsh (1976). The proof of Berge's extension of Tutte's theorem in Proposition 4.1.8 is based in Anderson's (1971) proof of Tutte's theorem and Brualdi's (1971) proof of Proposition 4.1.11 for infinite graphs, but the method of proof was first used by Gallai (1963). Proposition 4.1.11 was rediscovered by Las Vergnas (1973). The proof given here is based on McCarthy (1975); the first half of the proof establishes a result of Lovasz (1970). Transversal matroids were rediscovered by Mirsky and Perfect (1967). The proof of Proposition 4.2.6 is based in Brualdi (1971). Theorem 4.2.7 was discovered by Perfect (1969a), but the proof given here is based on an unpublished manuscript of Brualdi (1967). The proof of Proposition 4.3.2 is from Mirsky and Perfect (1967). For different approaches to many of the theorems proved here and for other related results, see Welsh (1976). For a more comprehensive treatment of transversal theory, see Mirsky (1971).

Exercises

[A general reference for most of the results contained in the exercises is Mirsky (1971)]

4.1.   Let $(A_1,..,A_n)$ be a family of subsets of a set S and let $p_1,...,p_n$ be non-negative integers. Then there exists a family $(X_1,...,X_n)$ of pairwise disjoint sets with $X_1 \subseteq A_i$ and $|X_i| = p_i$ for $i = 1,...,n$ if and only if

$$|A(J)| \geqslant \sum_{i \in J} p_i \quad (J \subseteq \{1,...,n\}).$$

(P.R. Halmos and H.E. Vaughan)

4.2.   Let $\mathscr{A}(I) = (A_i : i \in I)$ be a finite family of subsets of S, and let k be a positive integer. Then there is a partition $I_1,...,I_k$ of I such that $\mathscr{A}(I_j)$ has a transversal for $1 \leqslant j \leqslant k$ if and only if

$$k|A(J)| \geqslant |J| \quad (J \subseteq I).$$

(R. Rado)

4.3.   Let $\mathscr{A}(I) = (A_i : i \in I)$ be a finite family of subsets of S. Let $I' \subseteq I$ and let $S' \subseteq S$. Then the following are equivalent:

(i) $\mathscr{A}(I')$ has a transversal, and S' is a partial transversal of $\mathscr{A}(I)$.

(ii) There exists S'' with $S' \subseteq S'' \subseteq S$ and there exists I'' with $I' \subseteq I'' \subseteq I$ such that S'' is a transversal if $\mathscr{A}(I'')$.

(A.L. Dulmage and N.S. Mendelsohn)

4.4.   Let $\mathscr{A} = (A_1,...,A_n)$ be a family of subsets of S and let $r_1,...,r_m$ be non-negative integers less than or equal to n. Then $\mathscr{A}$ has m pairwise disjoint, partial transversals of cardinalities $r_1,...,r_m$ respectively if and only if

$$|A(J)| \geqslant \sum_{i \in J} (|J| - n + r_i)^+ \quad (J \subseteq I).$$

Here for a real number a we define a^+ to be a unless $a < 0$ in which case $a^+ = 0$.
(P.J. Higgins)

4.5.   A bipartite graph is *regular* of degree k if each vertex meets exactly k edges. Prove that the edges of a regular bipartite graph of degree k can be partitioned into k perfect matchings. (König 1931)

4.6.   An n by n matrix is *doubly stochastic* if its entries are non-negative numbers which
        sum up to 1 in each row and column. Prove that a doubly stochastic matrix can be
        written as $c_1 P_1 + \cdots + c_t P_t$ where $P_1, \ldots, P_t$ are permutation matrices and
        $c_1, \ldots, c_t$ are positive numbers summing up to 1. (G. Birkhoff)
4.7.   Let $\mathscr{A}(I) = (A_i : i \in I)$ and $\mathscr{B}(J) = (B_j : j \in J)$ be two finite families of a set S. Then
        the maximum integer p such that there exists a set T of cardinality p which is
        a partial transversal of both $\mathscr{A}(I)$ and $\mathscr{B}(J)$ equals

$$|I| + |J| + \min_{K,L} \{ |A(K) \cap B(L)| - |K| - |L| \}$$

        where the minimum is taken over all subsets K of I and L of J.
4.8.   Prove that a graph which is a tree has at most one perfect matching.
4.9.   Let $M_1(S), \ldots, M_k(S)$ be finite matroids with rank functions $r_1, \ldots, r_k$, respectively.
        Prove that there is a partition of S into sets $S_1, \ldots, S_k$ such that S_i is an
        independent set of $M_i(S)$ for $i = 1, \ldots, k$ if and only if

$$|X| \leqslant r_1(X) + \cdots + r_k(X)$$

        for all $X \subseteq S$. (Edmonds and Fulkerson 1965)
4.10.   Prove that the Fano matroid is not a transversal matroid.
4.11.   Prove that the duals of the polygon matroids of the complete graph K_5 and the
        complete bipartite graph $K_{3,3}$ are not transversal matroids.
4.12.   Let $\mathscr{A}(I) = (A_i : i \in I)$ and $\mathscr{B}(J) = (B_j : j \in J)$ be two finite families of subsets of the
        finite sets S. Show by example that the collection of subsets of S which are partial
        transversals of both $\mathscr{A}(I)$ and $\mathscr{B}(J)$ need not be the collection of independent sets
        of a matroid.

References

Aigner, M. and Dowling, T.A. (1971). Matching theory for combinatorial geometries. *Trans. Amer. Math. Soc.* **158**, 231–45.

Anderson, I. (1971). Perfect matchings of graphs. *J. Comb. Theory* **10**, 183–6.

Berge, C. Sur le couplage maximum d'un graphe. *C.R. Acad. Sciences (Paris)* **247**, 258–9.

Brualdi, R.A. (1970). Admissible mappings between dependence spaces. *Proc. London Math. Soc.* (3), **21**, 296–312.

Brualdi, R.A. (1971). Matchings in arbitrary graphs. *Proc. Cambridge Phil. Soc.* **69**, 401–7.

Edmonds, J. (1970). Submodular functions, matroids, and certain polyhedra, in *Proc. Int. Conf. on Combinatorics (Calgary)*, pp. 69–87. Gordon and Breach, New York.

Edmonds, J. and Fulkerson, D.R. (1965). Transversals and matroid partition. *J. Res. Nat. Bur. Stand.* **69B**, 147–53.

Ford, L.R., Jr. and Fulkerson, D.R. (1958). Network flows and systems of representatives. *Cand. J. Math.* **10**, 78–84.

Gallai, T. Neuer Beweiss eines Tutte' schen Stolzes. *Magyar Tud. Akad. Mat. Kutato Int. Kozl.* **8**, 135–9.

Hall, P. (1935). On representatives of subsets. *J. London Math. Soc.* **10**, 26–30.

Hoffman, A.J. and Kuhn, H.W. (1956). On systems of distinct representatives, linear inequalities and related systems. *Ann. Math. Studies* **38**, (*Princeton*), 199–206.

König, D. (1931). Graphen és matrixok. *Mat. Fiz. Lapote* **38**, 116–19.

Las Vergnas, M. (1973). Degree constrained subgraphs and matroid theory, in *Colloq. Math. Societatis Janos Bolyai* (Infinite and Finite Sets), Keszthely.

Lovasz, L. (1970). Subgraphs with prescribed valencies. *J. Comb. Theory* **8**, 391–416.

McCarthy, P.J. (1975). Matchings in graphs II. *Discrete Math.* **11**, 141–7.

Mirsky, L. (1971). *Transversal Theory*. Academic Press, London.

Mirsky, L. and Perfect, H. (1967). Applications of the notion of independence to combinatorial analysis. *J. Comb Theory* **2**, 327–57.

Ore, O. (1955). Graphs and matching theorems. *Duke Math. J.* **22**, 625–39.

Perfect, H. (1969a). Independence spaces and combinatorial problems. *Proc. London Math. Soc.* **19**, 17–30.

Perfect, H. (1969b). A generalization of Rado's theorem on independent transversals. *Proc. Cambridge Phil. Soc.* **66**, 513–15.

Rado, R. (1949). A theorem on independence relations. *Quart. J. Math. (Oxford)* **20**, 95–104.

Tutte, W.T. (1947). The factorization of linear graphs. *J. London Math. Soc.* **22**, 107–11.

Welsh, D.J.A. (1976). *Matroid Theory*. Academic Press, London.

5

Transversal Matroids

RICHARD A. BRUALDI

5.1 Introduction

In this chapter we study more comprehensively the transversal matroids which were introduced in the preceding chapter. Recall that these matroids are defined as follows. Let $\mathscr{A}(I) = (A_i : i \in I)$ be a finite family of subsets of a finite set S. There is no loss of generality in taking the index set I to be $\{1, 2, \ldots, n\}$, and we write $\mathscr{A} = (A_1, A_2, \ldots, A_n)$. The *transversal matroid* $M_{\mathscr{A}}(S)$ *of the family* $\mathscr{A}$ *of subsets of* S is the matroid on S whose collection $\mathscr{I}_{\mathscr{A}}$ of independent sets is the set of partial transversals of $\mathscr{A}$. We say that a matroid $M(S)$ is a *transversal matroid* provided there is some finite family $\mathscr{A}$ of subsets of S such that $M(S)$ coincides with $M_{\mathscr{A}}(S)$. The family $\mathscr{A}$ is then called a *presentation* of M. It is readily discovered that a transversal matroid has in general many presentations. As an example, take S to be $\{a, b, c, d, e\}$ and take $M(S)$ to be the 3-uniform matroid on S whose independent sets are all subsets of S with at most three elements. Then with $A_1 = \{a, b, c\}$, $A_2 = \{a, b, d\}$, and $A_3 = \{a, b, e\}$, (A_1, A_2, A_3) is a presentation of $M(S)$. But then so is (X_1, X_2, X_3) whenever $A_i \subseteq X_i \subseteq S$ for $i = 1, 2, 3$. While a presentation of a transversal matroid is not uniquely determined, we show in the next section that a transversal matroid of rank k has a unique 'maximal presentation' $(M_1, M_2, \ldots, M_k)$. For the matroid $M(S)$ above, the maximal presentation is (S, S, S).

We begin with some general properties of presentations and first observe that a transversal matroid of rank k can be presented by k (but clearly no fewer) sets.

5.1.1. Lemma. *Let $M(S)$ be a transversal matroid of rank k with presentation $\mathscr{A} = (A_1, A_2, \ldots, A_n)$. Let $\{i_1, i_2, \ldots, i_k\}$ be any subset of $\{1, \ldots, n\}$ of cardinality k such that the subfamily $\mathscr{A}' = (A_{i_1}, A_{i_2}, \ldots, A_{i_k})$ of $\mathscr{A}$ has a transversal. Then $\mathscr{A}'$ is also a presentation of $M(S)$.*

Proof. The proof is very similar to the proof of Theorem 4.2.7 and consequently we omit it.   □

Let F be a flat (or more generally a set of elements) of the matroid $M(S)$. Then F is *cyclic* provided F has no isthmuses, that is, provided each element of F is in a circuit which is entirely contained in F. In terms of the rank function, F is cyclic if and only if $r(F\backslash\{a\}) = r(F)$ for all $a \in F$. Let B be a basis of F and for $x \in F\backslash B$ let C_x be the unique circuit with $x \in C_x \subseteq B \cup \{x\}$. Then it is straightforward to check that F is cyclic if and only if $B \subseteq \cup_{x \in F\backslash B} C_x$. We say that the *matroid $M(S)$ is cyclic* when S is cyclic. The following result of Mason (1970) and Brualdi and Mason (1972) is a kind of converse of Lemma 5.1.1 for cyclic matroids.

5.1.2. Lemma. *Let $M(S)$ be a cyclic transversal matroid of rank k and let $(A_1, A_2, \ldots, A_n)$ be a presentation of $M(S)$. Then exactly k of the sets $A_1, A_2, \ldots, A_n$ are non-empty.*

Proof. Let $B = \{b_1, \ldots, b_k\}$ be a basis of $M(S)$. We may suppose without loss of generality that $b_i \in A_i\,(i = 1, \ldots, k)$. We then need to show that $A_{k+1} = \cdots = A_n = \varnothing$. Since $M(S)$ has rank k, it follows that $A_i \subseteq B$ for $i = k+1, \ldots, n$. Suppose $A_{k+1} \neq \varnothing$, and let $b_j \in A_{k+1}$. Since $M(S)$ is cyclic, there exists $x \in S\backslash B$ and a circuit C_x such that $\{x, b_j\} \subseteq C_x \subseteq B \cup \{x\}$. Then $(B \cup \{x\})\backslash\{b_j\}$ is a basis of $M(S)$ and by Lemma (5.1.1) is a transversal of $(A_1, \ldots, A_k)$. Since $b_j \in A_{k+1}, B \cup \{x\}$ is a transversal of $(A_1, \ldots, A_k, A_{k+1})$, which contradicts the fact that B is a basis of $M(S)$. Hence $A_{k+1} = \varnothing$, and similarly $A_i = \varnothing$ for all $i = k+1, \ldots, n$.   □

5.1.3. Corollary. *Let $M(S)$ be a transversal matroid with presentation $(A_1, A_2, \ldots, A_n)$. For F a cyclic flat of rank k,*

$$|\{i : F \cap A_i \neq \varnothing\}| = k.$$

Proof. We need only apply the preceding lemma to the restriction $M(F)$, a transversal matroid of rank k presented by $(A_1 \cap F, A_2 \cap F, \ldots, A_n \cap F)$.   □

The next lemma was first observed by Brualdi and Dinolt (1972). It exhibits an important connection between a transversal matroid and the sets in a presentation.

5.1.4. Lemma. *Let $M(S)$ be a transversal matroid with presentation $(A_1, A_2, \ldots, A_n)$. Then $S\backslash A_i$ is a flat for $i = 1, \ldots, n$.*

Proof. It suffices to prove that $S\backslash A_1$ is a flat. This is surely the case when $A_1 = \varnothing$, so we suppose $A_1 \neq \varnothing$. Let B be a basis of the restriction

$M(S\backslash A_1)$, a transversal matroid with presentation $(A_2\backslash A_1,\ldots,A_n\backslash A_1)$. Let $x\in A_1$. Since B is a partial transversal of $(A_2\backslash A_1,\ldots,A_n\backslash A_1)$, it follows that $B\cup\{x\}$ is a partial transversal of $(A_1,A_2,\ldots,A_n)$. Hence for all $x\in A_1$, $(S\backslash A_1)\cup\{x\}$ has a larger rank than $S\backslash A_1$, and we conclude $S\backslash A_1$ is a flat.   □

In the next section we consider more specific properties of presentations of transversal matroids which lead to a characterization of this interesting class of matroids.

5.2. Presentations

Because of Lemma 5.1.1 we restrict our attention to presentations $(A_1,\ldots,A_k)$ of transversal matroids of rank k. The first three results are due to Bondy and Welsh (1971).

5.2.1. Lemma. Let $\mathscr{A}=(A_1,\ldots,A_k)$ be a presentation of the transversal matroid $M(S)$ of rank k. Let P be a transversal of $(A_2,\ldots,A_k)$ such that $P\cap A_1$ has minimum cardinality. Then $\mathscr{A}'=(A_1\backslash P, A_2,\ldots,A_k)$ is also a presentation of $M(S)$.

Proof. We note that the cardinality of $P\cap A_1$ equals $(k-1)-t$ where t is the rank of the transversal matroid with presentation $(A_2\backslash A_1,\ldots,A_k\backslash A_1)$. Since every transversal of $\mathscr{A}'$ is a transversal of $\mathscr{A}$, it suffices to prove that every transversal of $\mathscr{A}$ [basis of $M(S)$] is a transversal of $\mathscr{A}'$. Let $B=\{b_1,\ldots,b_k\}$ be a transversal of $\mathscr{A}$ with $b_i\in A_i$ for $i=1,\ldots,k$. Let $P=\{p_2,\ldots,p_k\}$ with $p_j\in A_j$ for $j=2,\ldots,k$, and set X equal to $P\cap A_1$. If $b_1\in A_1\backslash P$, then B is a transversal of $\mathscr{A}'$. Hence we may assume that $b_1\in P$. Without loss of generality let $b_1=p_2$, so that in particular $p_2\in P\cap A_1$. If $b_2\in A_1-P$, then $\{b_2,b_1=p_2,b_3,\ldots,b_k\}=B$ is a transversal of $\mathscr{A}'$. Thus we may assume that $b_2\notin A_1-P$. Suppose $b_2\notin P$. Then it follows that $P'=\{b_2,p_3,\ldots,p_k\}$ is a transversal of $(A_2,\ldots,A_k)$ and hence $|P\cap A_1|\leqslant|P'\cap A_1|$. Since $p_2\in P\cap A_1$, we conclude that $b_2\in A_1$ and hence that $b_2\in A_1-P$, a contradiction. We conclude that $b_2\in P$, and without loss of generality we take $b_2=p_3$. The above argument may be repeated. Since $|B|>|P|$, eventually we determine a j such that with properly chosen notation, $\{b_j,b_1=p_2,b_2=p_3,\ldots,b_{j-1}=p_j,b_{j+1},\ldots,b_k\}=B$ is a transversal of $\mathscr{A}'$. □

Consider a transversal matroid $M(S)$ of rank k with presentation $(A_1,A_2,\ldots,A_k)$. Then clearly $A_i\neq\varnothing (i=1,\ldots,k)$ and using Lemma 5.1.4 we conclude that each $S\backslash A_i$ is a flat with rank at most $k-1$. Suppose, for instance, $S\backslash A_1$ has rank equal to $k-1$. Then $S\backslash A_1$ is a hyperplane, equivalently A_1 is a cocircuit, and hence $(A_1',A_2,\ldots,A_k)$ is not a presentation of $M(S)$ for any proper subset A_1' of A_1. We now show that $M(S)$ has a presentation of k cocircuits, which then is a *minimal presentation* in the sense that no element can be removed from any set of the presentation. But first we remark that in a presentation $(A_1, A_2, \ldots, A_k)$ of $M(S)$, if two sets are equal, say $A_1 = A_2$, then $(A_1 \backslash \{a\}, A_2, \ldots, A_k)$ is a presentation for any $a \in A_1$. It follows that the cocircuits in any minimal presentation are necessarily distinct.

5.2.2. Theorem. *Let* $\mathscr{A} = (A_1, A_2, \ldots, A_k)$ *be a presentation of the transversal matroid* $M(S)$ *of rank* k. *Then there exist distinct cocircuits* $D_1, D_2, \ldots, D_k$ *such that* $D_i \subseteq A_i$ $(i = 1, \ldots, k)$ *and* $\mathscr{D} = (D_1, D_2, \ldots, D_k)$ *is a presentation of* $M(S)$.

Proof. We refer to Lemma 5.2.1. We have that $\mathscr{A}' = (A_1 \backslash P, A_2, \ldots, A_k)$ is a presentation of $M(S)$ and P is a transversal of $(A_2, \ldots, A_k)$ which is disjoint from $A_1 \backslash P$. Hence $P \subseteq S \backslash (A_1 \backslash P)$ and it follows that $S \backslash (A_1 \backslash P)$ is a flat of rank equal to $k - 1$, that is, a hyperplane. Hence $A_1 \backslash P$ is a cocircuit of $M(S)$. Applying Lemma 5.2.1 to $A_2, \ldots, A_k$ in turn, we arrive at a presentation $(D_1, D_2, \ldots, D_k)$ where $D_i \subseteq A_i$ and D_i is a cocircuit $(i = 1, 2, \ldots, k)$. These cocircuits are necessarily distinct and the theorem follows. $\square$

We note that a transversal matroid may have many different minimal presentations. For instance, let $M(S)$ be the 3-uniform matroid on $S = \{a, b, c, d, e\}$. Then

$$(\{a, b, c\}, \{a, b, d\}, \{a, b, e\})$$

and

$$(\{b, c, a\}, \{b, c, d\}, \{b, c, e\})$$

are both minimal presentations of $M(S)$. We have already remarked that a transversal matroid has a unique maximal presentation (apart from the ordering of the sets). As a step towards proving this fact, we determine when a set in a presentation may be enlarged without changing the matroid presented.

5.2.3. Proposition. *Let* $\mathscr{A} = (A_1, A_2, \ldots, A_k)$ *be a presentation of the transversal matroid* $M(S)$ *of rank* k, *and let* $a \in S \backslash A_1$. *Then* $\mathscr{A}' = (A_1 \cup \{a\}, A_2, \ldots, A_k)$ *is also a presentation of* $M(S)$ *if and only if* a *is an isthmus of the restriction* $M(S \backslash A_1)$.

Proof. The matroid $M(S \backslash A_1)$ is a transversal matroid with presentation $(A_2 \backslash A_1, \ldots, A_k \backslash A_1)$. First suppose that $\mathscr{A}'$ is a presentation of $M(S)$, and let B be a basis of $M(S \backslash A_1)$. Then B is a partial transversal of $(A_2 \backslash A_1, \ldots, A_k \backslash A_1)$ and hence of $(A_2, \ldots, A_k)$. Thus $B \cup \{a\}$ is a partial transversal of $\mathscr{A}'$ and hence, by our assumption, of $\mathscr{A}$. Since $(B \cup \{a\}) \cap A_1 = \varnothing$, $B \cup \{a\}$ is a partial transversal of $(A_2 \backslash A_1, \ldots, A_k \backslash A_1)$. Since B is a basis of $M(S \backslash A_1)$, it follows that $a \in B$. Since B is an arbitrary basis of $M(S \setminus A_1)$, it follows that a is an isthmus of $M(S \setminus A_1)$.

Conversely, suppose a is an isthmus of $M(S \setminus A_1)$. It then follows that every transversal P of $(A_2, \ldots, A_k)$ has a non-empty intersection with $A_1 \cup \{a\}$. We choose a P so that $P \cap (A_1 \cup \{a\})$ has minimum cardinality as follows. We take a maximum partial transversal P' of $(A_2 \setminus A_1, \ldots, A_k \setminus A_1)$. Since a is an isthmus of $M(S \setminus A_1)$, $a \in P'$. Then we observe that P' is a partial transversal of $(A_2, \ldots, A_k)$ and hence can be enlarged to a transversal P of $(A_2, \ldots, A_k)$. This P has minimum cardinality intersection with $A_1 \cup \{a\}$. We now apply Lemma 5.2.1 to the transversal matroid $M'(S)$ with presentation $\mathscr{A}'$. Since $a \in P$, we conclude that $(A_1 \setminus P, A_2, \ldots, A_k)$ is a presentation of $M'(S)$ and hence that $\mathscr{A} = (A_1, A_2, \ldots, A_k)$ is a presentation of $M'(S)$. Hence $M(S) = M'(S)$, and it follows that $\mathscr{A}'$ is a presentation of $M(S)$. $\qquad \square$

We say that the presentation $(A_1, \ldots, A_k)$ of the transversal matroid $M(S)$ of rank k is a *maximal presentation* provided every presentation $(A'_1, \ldots, A'_k)$ of $M(S)$ with $A_i \subseteq A'_i$ for $i = 1, \ldots, k$ satisfies $A_i = A'_1 (i = 1, \ldots, k)$. A transversal matroid always has at least one maximal presentation.

5.2.4. Corollary. *Let* $(A_1, \ldots, A_k)$ *be a maximal presentation of the transversal matroid $M(S)$ of rank k. Then $S \setminus A_i$ is a cyclic flat for $i = 1, \ldots, k$.*

Proof. This is an immediate consequence of Lemma 5.1.4 and Proposition 5.2.3. $\qquad \square$

In Theorem 5.2.6 we shall verify an algorithm for determining a maximal presentation of a transversal matroid. A consequence of this algorithm will be the uniqueness of a maximal presentation. The first step of the algorithm is contained in the next lemma.

5.2.5. Lemma. *Let* $\mathscr{A} = (A_1, \ldots, A_k)$ *be a presentation of the transversal matroid $M(S)$ of rank k. Let $F_1, \ldots, F_t$ be the distinct cyclic hyperplanes of $M(S)$. Then after possibly renumbering the sets in $\mathscr{A}$, we have*

$$A_i = S \setminus F_i \quad (i = 1, \ldots, t)$$
$$A_j \neq S \setminus F_i \quad (j = t+1, \ldots, k; i = 1, \ldots, t).$$

If $(M_1, \ldots, M_k)$ is a maximal presentation of $M(S)$ and F is a hyperplane different from $F_1, \ldots, F_t$, then $M_i \neq S \setminus F$ for $i = 1, \ldots, k$.

Proof. Since the $S \setminus F_i$ are cocircuits, it suffices to consider only maximal presentations $(M_1, \ldots, M_k)$. Consider the cyclic hyperplane F_1. It follows from Corollary 5.1.3 that F_1 has a non-empty intersection with exactly $k - 1$ of the sets $M_1, \ldots, M_k$. Relabeling, if necessary, we may suppose that $F_1 \cap M_1 = \varnothing$ so that $M_1 \subseteq S \setminus F_1$. Since $S \setminus F_1$ is a cocircuit, we now conclude that

$M_1 = S \backslash F_1$. Continuing like this, we obtain that $M_i = S \backslash F_i$ for $i = 1, \ldots, t$. Suppose there were a j with $t < j \leqslant k$ such that $M_j = S \backslash F_i$ with $1 \leqslant i \leqslant t$. Then $M_i = M_j$. If a is any element of M_j, we may replace M_j by $M_j \backslash \{a\}$ and still have a presentation of $M(S)$. Since M_j is a cocircuit, this is a contradiction. Hence $M_j \neq S \backslash F_i$ for $t < j \leqslant k$ and $1 \leqslant i \leqslant t$. Finally, let F be any hyperplane different from $F_1, \ldots, F_t$. Then F is not cyclic, and it follows from Corollary 5.2.4 that $M_i \neq S \backslash F$ for $i = 1, \ldots, k$. $\square$

Let $M(S)$ be a matroid of rank k with lattice of flats $\mathscr{L}$. Let $\mathscr{F}$ be the subset of $\mathscr{L}$ consisting of the cyclic flats of $M(S)$. The join in $\mathscr{A}$ of two cyclic flats is again a cyclic flat and hence $\mathscr{F}$ is a join subsemilattice of $\mathscr{L}$. For a cyclic flat F we let $\mathscr{K}(F)$ denote the set of all cyclic flats which properly contain F. For a real number a, let a^+ equal a if $a > 0$ and 0, otherwise. We define an integer-valued function τ on $\mathscr{F}$ recursively as follows: we set $\tau(S) = 0$ if S is a cyclic flat of $M(S)$ [otherwise $\tau(S)$ is undefined]. For $j = 1, \ldots, k$, let $\mathscr{F}_j$ be the set of cyclic flats of rank $k - j$. For $j = 1, \ldots, k$, and each $F \in \mathscr{F}_j$, let

$$\tau(F) = \left[k - r(F) - \sum_{K \in \mathscr{K}(F)} \tau(K) \right]^+$$
$$= \left[j - \sum_{K \in \mathscr{K}(F)} \tau(K) \right]^+.$$

It follows from this definition that if $F \in \mathscr{F}_1$, that is, F is a cyclic hyperplane, then $\tau(F) = 1$. If $\in \mathscr{F}_2$ so that F is a cyclic flat of rank $k - 2$, $\tau(F) = 2, 1$, or 0 according as there are 0, 1, or more than 1 cyclic hyperplanes containing F. In general if $F \in \mathscr{F}_j$, $0 \leqslant \tau(F) \leqslant j$. We let $\mathscr{B} = (F_1, \ldots, F_n)$ be the family of cyclic flats defined by the property that each cyclic flat F of $M(S)$ occurs $\tau(F)$ times in $\mathscr{B}$. We call $\mathscr{B}$ the *distinguished family of cyclic flats* of $M(S)$. We note that since the closure $\varnothing$ of the empty set $\varnothing$ is a cyclic flat of rank 0 which is properly contained in every cyclic flat of rank at least 1, the number n of flats in $\mathscr{B}$ is at least k.

The significance of the family $\mathscr{B}$ is contained in the following result of Brualdi and Dinolt (1972).

5.2.6. Theorem. *Let $M(S)$ be a transversal matroid of rank k with distinguished family of cyclic flats $\mathscr{B} = (F_1, \ldots, F_n)$. Then*

(i) $\sum_{K \in \mathscr{K}(F)} \tau(K) \leqslant k - r(F)$ *for each cyclic flat F,*

(ii) $n = k$,

(iii) $(S/F_1, \ldots, S/F_k)$ *is, apart from order, the unique maximal presentation of $M(S)$.*

Proof. Let $\mathscr{M} = (M_1, \ldots, M_k)$ be any maximal presentation of $M(S)$. We prove by induction on j that if $F \in \mathscr{F}_j$, then (i) holds and S/F occurs exactly $\tau(F)$ times in $\mathcal{M}$. Since the S/M_i are cyclic flats, the theorem will follow. For $j = 1$, the conclusion holds by Lemma 5.2.5 and the definition of τ. We assume $j > 1$ and proceed by induction. Let $F \in \mathcal{F}_j$ so that F is a cyclic flat of rank $k - j$. It follows from Corollary 5.1.3 that there exists $I \subseteq \{1, \ldots, k\}$ with $|I| = k - j$ such that $F \cap M_i \neq \varnothing$ for $i \in I$ and $F \subseteq S/M_i$ for $i \notin I$. For each i with $F \subsetneqq S/M_i$, S/M_i is a cyclic flat of rank at least $k - (j - 1)$ which properly contains F so that $S/M_i \in \mathcal{K}(F)$. By the inductive assumption for each $K \in \mathcal{K}(F)$, S/K occurs exactly $\tau(K)$ times in $\mathcal{M}$. It follows that

$$\sum_{K \in \mathcal{K}(F)} \tau(K) = |\{i : F \subsetneqq S \backslash M_i\}| \leqslant j = k - r(F)$$

and that exactly $\tau(F)$ of the sets $S \backslash M_i$ equal F. Hence the induction is complete. It follows from the definition of τ and (i) applied to $\bar{\varnothing}$ that $n = k$.   □

We note that Theorem 5.2.6 furnishes an algorithm for obtaining a maximal presentation of a transversal matroid. The uniqueness of a maximal presentation is a consequence of this algorithm. This uniqueness was first proved by Mason (1970).

Theorem 5.2.6 furnishes a necessary condition for a matroid of rank k to be a transversal matroid, namely

$$k - r(F) - \sum_{K \in \mathcal{K}(F)} \tau(K) \geqslant 0 \quad \text{for all cyclic flats } F. \tag{5.1}$$

But (5.1) is not a sufficient condition as the following example shows.

5.2.7. Example. Let $M(S)$ be the rank 3 matroid on $S = \{1, 2, 3, 4, 5, 6, 7\}$ whose bases are all the 3-element subsets of S except $F_1 = \{1, 2, 3\}$, $F_2 = \{1, 4, 5\}$, and $F_3 = \{1, 6, 7\}$. Then $M(S)$ is the affine matroid pictured in Figure 5.1. The set $\mathcal{F}$ of cyclic flats of $M(S)$ is $\{\varnothing, F_1, F_2, F_3, S\}$, and the distinguished family of cyclic flats of $M(S)$ is $\mathcal{B} = (F_1, F_2, F_3)$, and (5.1) holds. But it is readily established that $M(S)$ is not a transversal matroid [see also (5.2) below].

We now determine necessary and sufficient conditions on the distinguished family $\mathcal{B}$ of cyclic flats in order that a matroid be a transversal matroid. As a first step we prove the following.

Figure 5.1. A non-transversal matroid.

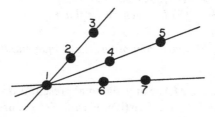

5.2.8. Lemma. *Let $M(S)$ be a matroid of rank k, and let $M'(S)$ be the transversal matroid with presentation $\mathscr{A} = (A_1, A_2, \ldots, A_k)$. Then every independent set of $M(S)$ is also an independent set of $M'(S)$ if and only if*

$$r\left(\bigcap_{i \in I}(S \backslash A_i)\right) \leqslant k - |I|, \quad \text{for all} \quad I \subseteq \{1, \ldots, k\}. \tag{5.2}$$

Proof. First suppose (5.2) holds and let B be a basis of $M(S)$. We show that B is a transversal of $\mathscr{A}$, equivalently that B is a transversal of $(A_1 \cap B, \ldots, A_k \cap B)$. Suppose that for some $I \subseteq \{1, \ldots, k\}$,

$$\left|\left(\bigcup_{i \in I} A_i\right) \cap B\right| < |I|.$$

Then

$$r\left(\bigcap_{i \in I}(S \backslash A_i)\right) \geqslant \left|B \cap \left(S \backslash \bigcup_{i \in I} A_i\right)\right| \geqslant k - |I| + 1,$$

contradicting (5.2). Hence $|(\bigcup_{i \in I} A_i) \cap B| \geqslant |I|$ for all $I \subseteq \{1, \ldots, k\}$, and by Hall's theorem (Proposition 4.1.4), $(A_1 \cap B, \ldots, A_k \cap B)$ has a transversal. Since $M(S)$ has rank k, $|B| = k$ and B is that transversal. It follows that B is a basis of $M'(S)$.

Now suppose every basis of $M(S)$ is a basis of $M'(S)$. Then for each basis B of $M(S)$, B is a transversal of $(A_1, \ldots, A_k)$ and hence $(A_1 \cap B, \ldots, A_k \cap B)$ has a transversal. By Hall's theorem again,

$$\left|\left(\bigcup_{i \in I} A_i\right) \cap B\right| \geqslant |I|$$

for each $I \subseteq \{1, \ldots, k\}$ and each basis B of $M(S)$, and (5.2) follows. $\qquad\square$

The following characterization of transversal matroids is due to Brualdi and Dinolt (1972).

5.2.9. Proposition. *Let $M(S)$ be a matroid of rank k and let $\mathscr{B} = (F_1, \ldots, F_n)$ be its distinguished family of cyclic flats. Then $M(S)$ is a transversal matroid if and only if*

$$r\left(\bigcap_{i \in I} F_i\right) \leqslant k - |I|, \quad \text{for all} \quad I \subseteq \{1, \ldots, n\}. \tag{5.3}$$

Proof. It follows from Theorem 5.2.6 that $M(S)$ is a transversal matroid if and only if $n = k$ and $M(S)$ is the transversal matroid $M'(S)$ of the family $\mathscr{A} = (S \backslash F_1, \ldots, S \backslash F_n)$. Combining this with Lemma 5.2.8 we conclude that if $M(S)$ is a transversal matroid, then 5.3 holds.

Now suppose (5.3) holds. By taking $I = \{1, \ldots, n\}$ in (5.3) we see that $n \leqslant k$. Since it is always the case that $n \geqslant k$, we conclude $n = k$. It follows from Lemma

5.2.8 that every independent set of $M(S)$ is also an independent set of $M'(S)$. Hence to complete the proof we need only show that every transversal of $\mathscr{A}$ is an independent set of $M(S)$. Suppose to the contrary that T is a transversal of $\mathscr{A}$ but T is not an independent set of $M(S)$. Then T contains a circuit, and we let F be the cyclic flat which is spanned by the union D of the circuits contained in T. Let $r(F) = k - j$ where $j \geqslant 1$. Then

$$k - j = r(D) < |D|, \quad |T \backslash D| \leqslant j - 1.$$

Let $I = \{i : F \subseteq F_i\}$. Then $F \cap (S \backslash F_i) = \varnothing$ for $i \in I$, and it follows from the definition of $\mathscr{B}$ that $|I| \geqslant j$. From (5.3) we get

$$k - j = r(F) \leqslant r\left(\bigcap_{i \in I} F_i\right) \leqslant k - |I|.$$

Hence $|I| \leqslant j$, so that we now have $|I| = j$. Thus F intersects only $k - j$ sets of the family $\mathscr{A}$. Since $D \subseteq F, D$ intersects at most $k - j$ sets of $\mathscr{A}$. Since $|D| > k - j$ and since D, by virtue of being a subset of the transversal T of $\mathscr{A}$, is a partial transversal of $\mathscr{A}$, we have a contradiction. Hence every transversal T of $\mathscr{A}$ is an independent set of $M(S)$, and the proposition follows.   □

5.2.10. Remark. The characterization of transversal matroids given in Proposition 5.2.9 is readily seen to be equivalent to the following. Define an integer-valued function τ' on the set $\mathscr{F}$ of cyclic flats of a matroid $M(S)$ of rank k by

$$\tau'(F) = k - r(F) - \sum_{K \in \mathscr{K}(F)} \tau'(K).$$

Let $\mathscr{B}' = (F'_1, \ldots, F'_m)$ be the family of cyclic flats whereby each cyclic flat F of $M(S)$ occurs exactly $[\tau'(F)]^+$ times in $\mathscr{B}'$. If $M(S)$ is a transversal matroid, then $\tau(F) = \tau'(F)$ for each cyclic flat F, and $\mathscr{B} = \mathscr{B}'$. In general, $M(S)$ is a transversal matroid if and only if

$$r\left(\bigcap_{i \in I} F'_i\right) \leqslant k - |I| \quad \text{for all} \quad I \subseteq \{1, \ldots, m\}.$$

There is a characterization of transversal matroids due to Mason (1970) which was the first characterization discovered. His characterization involves all the cyclic *sets* of a matroid and as a result is more difficult to apply.

The following result of Brualdi and Dinolt (1972) characterizes all presentations of a transversal matroid in terms of the maximal presentation which can be found by Theorem 5.2.6.

5.2.11. Proposition. *Let $M(S)$ be a transversal matroid of rank k with maximal presentation $(M_1, \ldots, M_k)$. Let $\mathscr{A} = (A_1, \ldots, A_k)$ be a family of sets with $A_i \subseteq M_i$*

for $i = 1, \ldots, k$. *Then* $\mathscr{A}$ *is a presentation of* $M(S)$ *if and only if*

$$r\left(\bigcap_{i \in I}(S \backslash A_i)\right) \leqslant k - |I| \quad \text{for all} \quad I \subseteq \{1, \ldots, k\}.$$

Proof. We first note that since the maximal presentation is unique, the assumption that $A_i \subseteq M_i$ for $i = 1, \ldots, k$ is without loss of generality. The proposition now is an immediate consequence of Lemma 5.2.8. $\square$

The following characterization of transversal matroids is due to Ingleton (1975).

5.2.12. Proposition. *Let* $M(S)$ *be a matroid of rank* k. *Then* $M(S)$ *is a transversal matroid if and only if there exists a family* $(H_1, \ldots, H_k)$ *of hyperplanes such that*

$$r\left(\bigcap_{i \in I}H_i\right) \leqslant k - |I|, \quad \text{for all} \quad I \subseteq \{1, \ldots, k\}, \tag{5.4}$$

and

for each circuit C *there exists* $J \subseteq \{1, \ldots, k\}$ *with* $|J| = |C| - 1$

such that $C \subseteq \bigcap_{i \in J}H_i$. $\tag{5.5}$

When (5.4) *and* (5.5) *are satisfied,* $(S \backslash H_1, \ldots, S \backslash H_k)$ *is a presentation of* $M(S)$ *consisting of* k *cocircuits, a minimal presentation.*

Proof. First suppose that $M(S)$ is a transversal matroid. By Theorem 5.2.2 there is a family $\mathscr{D} = (D_1, \ldots, D_k)$ consisting of k distinct cocircuits such that $\mathscr{D}$ is a presentation of $M(S)$. Let $H_i = S \backslash D_i$ for $i = 1, \ldots, k$. Then $(H_1, \ldots, H_k)$ is a family of k hyperplanes, and it follows from Proposition 5.2.11 that (5.4) is satisfied and from Corollary 5.1.3 that (5.5) is satisfied.

Now suppose there is a family $(H_1, \ldots, H_k)$ of hyperplanes satisfying (5.4) and (5.5), and let $M'(S)$ be the transversal matroid with presentation $\mathscr{A} = (S \backslash H_1, \ldots, S \backslash H_k)$. It follows from Lemma 5.2.8 that every independent set of $M(S)$ is also an independent set of $M'(S)$. Suppose $M(S)$ were different from $M'(S)$. Then there exists a circuit C of $M(S)$ which is an independent set in $M'(S)$, that is, C is a partial transversal of $\mathscr{A}$. This contradicts (5.5) and the proposition follows. $\square$

By Theorem 5.2.2 every transversal matroid has a presentation consisting of cocircuits. The following result of Brualdi and Dinolt (1972) determines the cardinalities of these cocircuits with reference to the maximal presentation.

5.2.13. Proposition. *Let* $M(S)$ *be a transversal matroid of rank* k *with maximal presentation* $(M_1, \ldots, M_k)$, *and let* $(D_1, \ldots, D_k)$ *be a presentation of* $M(S)$ *where* D_i

is a cocircuit and $D_i \subseteq M_i$ for $i = 1, \ldots, k$. Then

$$|D_i| = |M_i| - ((k-1) - r(S \backslash M_i))$$

for $i = 1, \ldots, k$.

Proof. It suffices to obtain the above identity for $i = 1$. From the hypotheses it follows that both $(D_1, D_2, \ldots, D_k)$ and $(M_1, D_2, \ldots, D_k)$ are presentations of $M(S)$. Hence by Proposition 5.2.3 each element of $M_1 \backslash D_1$ is an isthmus of the restriction $M(S \backslash D_1)$. Hence

$$k - 1 = r(S \backslash D_1) = r(S \backslash M_1) + |M_1 \backslash D|$$

from which the proposition follows.   □

From the preceding theorem we obtain the following result of Bondy (1972a).

5.2.14. Corollary. *With the notation of Proposition 5.2.13, $|D_i|$ is the maximum cardinality of the cocircuits contained in M_i for $i = 1, \ldots, k$. In particular the cardinalities of the cocircuits in the minimal presentations of a transversal matroid are uniquely determined.*

Proof. If D is a cocircuit contained in M_i, then

$$\begin{aligned} |D| &\leqslant |M_i| - (r(S \backslash D) - r(S \backslash M_i)) \\ &= |M_i| - ((k-1) - r(S \backslash M_i)). \end{aligned}$$

The result now follows from Proposition 5.2.13.   □

5.3. Duals of Transversal Matroids

The dual of a transversal matroid need not be a transversal matroid, and the purpose of this section is to identify those matroids which are the duals of transversal matroids. As an example, let $S = \{a, b, c, d, e, f\}$ and let $M(S)$ be the rank 4 matroid which is pictured affinely in Figure 5.2. The distinguished

Figure 5.2. A transversal matroid whose dual is not transversal.

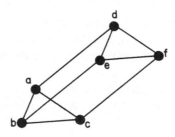

family of cyclic flats of $M(S)$ is $\mathscr{B} = (F_1, F_2, F_3, F_4)$ where $F_1 = \{b, c, e, f\}$, $F_2 = \{a, d, c, f\}, F_3 = \{a, b, d, e\}$, and $F_4 = \varnothing$. It is easily verified that $\mathscr{B}$ satisfies (5.3) and hence, by Proposition 5.2.9, $M(S)$ is a transversal matroid with maximal presentation (M_1, M_2, M_3, M_4) where $M_1 = \{a, d\}$, $M_2 = \{b, e\}, M_3 = \{c, f\}$, and $M_4 = S$. The matroid dual to $M(S)$ is the rank 2 matroid $M^*(S)$ which is pictured affinely in Figure 5.3. One easily checks that $M^*(S)$ is not a transversal matroid (or use Proposition 5.2.9).

Figure 5.3. The dual of the matroid in Figure 5.2.

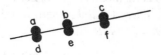

To characterize the duals of transversal matroids we define a new class of matroids by the use of directed graphs (*digraphs*, for short). This class of matroids was first introduced by Perfect (1968), and we shall follow the general development of Ingleton and Piff (1973) which leads to their identification as the duals of transversal matroids.

Let $\Gamma = \Gamma(S)$ be a *digraph* with a finite set S of *vertices*. Thus each *arc* of Γ is an ordered pair st of distinct vertices, and we generally denote the set of arcs of Γ by E. A *path* in Γ is a sequence $\gamma = (s_0, s_1, \ldots, s_k)$ of distinct vertices where $k \geqslant 0$ (at least one vertex) and where $s_{i-1} s_i$ is an arc for $i = 1, \ldots, k$. The arcs $s_0 s_1, s_1 s_2, \ldots, s_{k-1} s_k$ are called the *arcs of* γ and their number k is the *length of* γ. The *initial vertex* of γ is s_0 and the *terminal vertex* is s_k, and γ is said to *join* s_0 to s_k. We emphasize that we have allowed the path γ to have length 0, in which case it has no arcs and the terminal vertex is the same as the initial vertex. Two paths $\gamma = (s_0, s_1, \ldots, s_k)$ and $\delta = (t_0, t_1, \ldots, t_l)$ are *vertex disjoint* if $\{s_0, s_1, \ldots, s_k\} \cap \{t_0, t_1, \ldots, t_l\} = \varnothing$. The digraph Γ is said to be *bipartite* provided its vertex set S can be partitioned into two sets S_1 and S_2 such that each arc of Γ has initial vertex in S_1 and terminal vertex in S_2. We note that every bipartite graph $\Gamma(S_1, S_2)$ can be regarded as a bipartite digraph by directing each edge from S_1 to S_2.

Let A and B be subsets of the vertex set S of the digraph $\Gamma = \Gamma(S)$. Then A is said to be *linked into* B in Γ provided there exists a collection Θ of paths in Γ with the following properties:

(i) $|\Theta| = |A|$.

(ii) The paths in Θ are pairwise vertex disjoint.

(iii) The initial vertex of each path in Θ belongs to A while the terminal vertex belongs to B.

A collection Θ of paths satisfying (i), (ii), and (iii) is said to be a *linking of A into*

B. It follows that a linking of A into B defines in particular an injective mapping $f: A \to B$ by: for $a \in A$, let $f(a)$ equal the terminal vertex of the path in Θ whose initial vertex is a. When $|A| = |B|$, the mapping f is a bijection, and Θ is called a *linking of A onto B* and A is said to be *linked onto B in Γ*. For B fixed, the collection of subsets of S which are linked into B are the independent sets of a matroid of which B is a basis. Before proving this statement, we obtain the following result of Ingleton and Piff (1973). For each vertex s of Γ, let

$$A_s = \{t : st \quad \text{is an arc of} \quad \Gamma\} \cup \{s\}.$$

5.3.1. Lemma (*The fundamental linking lemma*). *Let $\Gamma = \Gamma(S)$ be a digraph, and let A and B be subsets of S. Then A is linked onto B in Γ if and only if $S \backslash A$ is a transversal of the family $\mathscr{A} = (A_s : s \in S \backslash B)$.*

Proof. First suppose that Θ is a linking of A onto B, so that in particular $|A| = |B|$. We define a function $\phi : S \backslash A \to S$ as follows (see Figure 5.4):

$$\phi(t) = \begin{cases} s, & \text{if } st \text{ is an arc of one of the paths in } \Theta, \\ t, & \text{otherwise.} \end{cases}$$

Figure 5.4. A path of Θ.

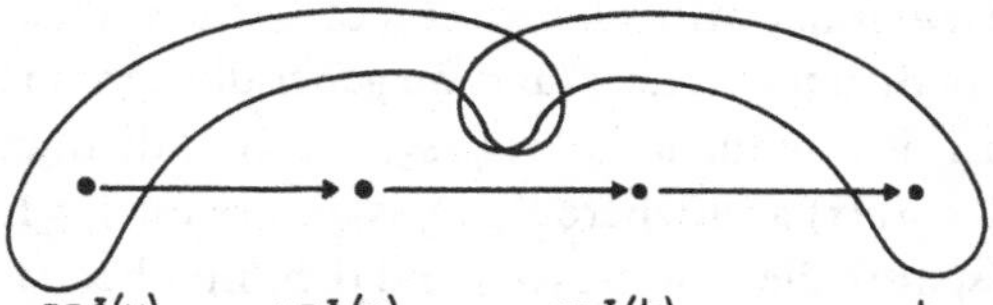

$$a = \phi(y) \qquad y = \phi(x) \qquad x = \phi(b) \qquad b$$

Suppose $t \in S \backslash A$. If t is not the terminal vertex of an arc of a path of Θ, then $t \notin B$ and $\phi(t) = t \in S \backslash B$. If t is the terminal vertex of an arc of a path of Θ, then $\phi(t) \neq t$ and $\phi(t) \in S \backslash B$. Hence $\phi : S \backslash A \to S \backslash B$. Since the paths in Θ are pairwise vertex disjoint, ϕ is an injection. Since $|A| = |B|$, we now conclude that $\phi : S \backslash A \to S \backslash B$ is a bijection. From the definition of the sets of the family $\mathscr{A}$, $t \in A_{\phi(t)}$ for each $t \in S \backslash A$ and it follows that $S \backslash A$ is a transversal of $\mathscr{A}$ (the mapping ϕ^{-1} defines a system of distinct representatives corresponding to the transversal $S \backslash A$).

Now suppose that $S \backslash A$ is a transversal of the family $\mathscr{A} = (A_s : s \in S \backslash B)$. Then there exists a bijection $\phi : S \backslash A \to S \backslash B$ such that $t \in \phi(t)$ for each $t \in S \backslash A$. Hence, by definition of the sets of $\mathscr{A}$, for each $t \in S \backslash A$ either $\phi(t) = t$ or $\phi(t)t$ is an arc of Γ. Now consider t with $t \in B \backslash A$, and consider the sequence $t, \phi(t), \phi^2(t), \ldots$ Since S is finite, either there exists an integer m so that $\phi^m(t) \in A$, in which case $\phi^{m+1}(t)$ is not defined, or there exist integers k and l with $k < l$ such that $\phi^k(t) = \phi^l(t)$. Since $t \in B$ and since the function values of ϕ are in $S \backslash B$, it follows that in the latter case $k \geqslant 1$. Suppose there were such a k and l and choose k to be minimal. Then $\phi^k(t) = \phi^l(t)$ implies $\phi(\phi^{k-1}(t)) = \phi(\phi^{l-1}(t))$. Since ϕ is a bijection, $\phi^{k-1}(t) = \phi^{l-1}(t)$ contradicting the minimality of k. It follows that there exists a positive integer m, such that $t, \phi(t), \ldots, \phi^m(t)$ are distinct vertices such that $\phi^m(t) \in A$ and $\gamma_t = (\phi^m(t), \ldots, \phi(t), t)$ is a path in Γ. Since ϕ is a bijection, the collection of paths $(\gamma_t : t \in B \setminus A)$ are pairwise vertex disjoint. For $t \in A \cap B$, let γ_t be the path (t) of length 0. Then the set of paths $\Theta = \{\gamma_t : t \in B\}$ is a linking of A onto B. $\qquad\square$

Let $\Gamma(S)$ be a digraph and let B be a subset of S. Let $\mathscr{I}_{\Gamma,B}(S)$ be the collection of all subsets A of S which are linked into B, and note that $B \in \mathscr{I}_{\Gamma,B}(S)$. That $\mathscr{I}_{\Gamma,B}(S)$ is the collection of independent sets of a matroid on S was first proved by Perfect (1968). That these matroids are precisely the duals of transversal matroids was discovered by Ingleton and Piff (1973).

5.3.2. Theorem. *Let $\Gamma = \Gamma(S)$ be a digraph and let $B \subseteq S$. Then $\mathscr{I}_{\Gamma,B}(S)$ is the collection of independent sets of a matroid $M_{\Gamma,B}(S)$. Moreover, a matroid $M(S)$ is the dual of a transversal matroid if and only if there is a digraph $\Gamma(S)$ and a subset B of S such that $M(S) = M_{\Gamma,B}(S)$.*

Proof. We have that $A \in \mathscr{I}_{\Gamma,B}(S)$ if and only if A is linked onto a subset B' of B, that is, by Lemma 5.3.1 if and only if $S \setminus A$ is a transversal of $(A_s : s \in S \setminus B')$. We use this fact to show that $A \in \mathscr{I}_{\Gamma,B}(S)$ if and only if $S \setminus A$ contains a transversal of $(A_s : s \in S \setminus B)$. If $S \setminus A$ is a transversal of $(A_s : s \in S \setminus B')$ for some $B' \subseteq B$, then clearly $S \setminus A$ contains a transversal of $(A_s : s \in S \setminus B)$. Now suppose that $S \setminus A$ contains a transversal of $(A_s : s \in S \setminus B)$. Then it follows from Lemma 5.3.1 that there exists $A' \subseteq S$ with $A \subseteq A'$ such that A' is linked onto B in Γ. Since $A \subseteq A'$, A is linked into B in Γ and hence $A \in \mathscr{I}_{\Gamma,B}(S)$. It follows that $\mathscr{I}_{\Gamma,B}(S)$ consists of those subsets A of S which are contained in the complement of some transversal of $(A_s : s \in S \setminus B)$. Hence $\mathscr{I}_{\Gamma,B}(S)$ is the collection of independent sets of the dual of the transversal matroid with presentation $(A_s : s \in S \setminus B)$.

Now suppose $M(S)$ is the dual of the transversal matroid $M^*(S)$ of rank k with presentation $\mathscr{A} = (A_i : 1 \leqslant i \leqslant k)$. Let $T = \{t_i : 1 \leqslant i \leqslant k\}$ be a transversal of $\mathscr{A}$ where $t_i \in A_i$ for $i = 1, \ldots, k$. Let $\Gamma = \Gamma(S)$ be the digraph with vertex set S whose arcs are those ordered pairs $t_i y$ where $y \in A_i \setminus \{t_i\}$ and $i = 1, \ldots, k$. Finally, let $B = S \setminus T$. We show that $M(S) = M_{\Gamma,B}(S)$. First, let $S \setminus U$ be a basis of $M(S)$. Then U is a transversal of $\mathscr{A}$ and we write $U = \{u_i : 1 \leqslant i \leqslant k\}$ where $u_i \in A_i$ for $i = 1, \ldots, k$. From the definition of Γ, we conclude that $t_i u_i$ is an arc of Γ whenever $t_i \neq u_i$. Let $J = \{i : t_i \neq u_i, i = 1, \ldots, k\}$. Corresponding to each $j \in J$ there is a uniquely determined path γ_j which joins t_j to $U \cap (S \setminus T)$ such that the paths in $\{\gamma_j : j \in J\}$ are pairwise vertex disjoint. The path γ_j is defined as follows. Determine the integer p such that $u_j = t_{j_1}, u_{j_1} = t_{j_2}, u_{j_2} = t_{j_3}, \ldots, u_{j_p} = t_{j_{p+1}}, u_{j_p} \in U \cap (S \setminus T)$. Then $\gamma_j = (t_j, t_{j_1}, t_{j_2}, t_{j_3}, \ldots, t_{j_{p+1}}, u_{j_p})$. For

$s \in (S \backslash T) \cap (S \backslash U)$, define γ_s to be the path (s) of length 0. Then the collection of paths $\{\gamma_j : j \in J\} \cup \{\gamma_s : s \in (S \backslash T) \cap (S \backslash U)\}$ is a collection of pairwise vertex disjoint paths which links $S \backslash U$ onto $S \backslash T$. It follows that $S \backslash U$ is also a basis of $M_{\Gamma, B}(S)$.

Now suppose $S \backslash U$ is a basis of $M_{\Gamma, B}(S)$. Then there is a collection Θ of pairwise vertex disjoint paths of Γ which link $S \backslash U$ onto $B = S \backslash T$. Let t_i be an element of $(S \backslash U) \cap T$ and let γ_i be the path in Θ with initial vertex t_i and terminal vertex in $S \backslash T$. Since the paths in Θ are pairwise vertex disjoint, each vertex of γ_i except for t_i belongs to U, indeed the terminal vertex belongs to $U \cap (S \backslash T)$ and the others belong to $U \cap T$. Let $u \in U$ and let u be a vertex of γ_i. Then there is a vertex $t_{\theta(u)} \in T$ immediately preceding u in γ_i. From the definition of Γ, it follows that $u \in A_{\theta(u)}$. If $u \in U$ is not a vertex of any path in Θ, then $u \in U \cap T$ so that $u = t_i$ for some i with $1 \leqslant i \leqslant k$, and we define $\theta(u) = i$. Since the paths in Θ are pairwise vertex disjoint, $\theta : U \to \{1, \ldots, k\}$ is a bijection with $u \in A_{\theta(u)}$ for all $u \in U$. Hence U is a transversal of $\mathscr{A}$ and hence a basis of $M^*(S)$. It follows that $M(S) = M_{\Gamma, B}(S)$.   □

A matroid of the form $M_{\Gamma, B}(S)$ where $\Gamma(S)$ is a digraph and $B \subseteq S$ has been called by Mason (1972) and others a *strict gammoid* while for $X \subseteq S$, the restriction $M_{\Gamma, B}(X)$ has been called a *gammoid*. The rank function of the matroid $M_{\Gamma, B}(S)$ or $M_{\Gamma, B}(X)$ is given by a classical theorem of Menger (see Theorem 5.3.3). Because of this historical connection and because of the artificiality of the term gammoid, we prefer to call these matroids *strict Menger matroids* and *Menger matroids*, respectively. Thus a Menger matroid is obtained by choosing a digraph $\Gamma(S)$ and two subsets X and B of the vertex set S; the independent sets are all those subsets of X which can be linked into B. When $X = S$, we obtain a strict Menger matroid having B as basis.

Suppose S_1 and S_2 are disjoint sets and $\Gamma = \Gamma(S_1, S_2)$ is a bipartite graph. As already remarked we may regard Γ as a bipartite digraph by directing all its edges from S_1 to S_2. The Menger matroid $M_{\Gamma, S_1}(S_2)$ is then the transversal matroid on S_2 corresponding to the bipartite graph $\Gamma(S_1, S_2)$. In particular, transversal matroids are Menger matroids. The strict Menger matroid $M_{\Gamma, S}(S_1 \cup S_2)$ is also a transversal matroid which has been termed a *principal transversal matroid* or *fundamental transversal matroid*. To see this let S_1' be a 'copy' of S_1 and let Γ' be the bipartite graph $\Gamma'(S_1, S_1' \cup S_2)$ obtained from $\Gamma(S_1, S_2)$ by adding an edge xx' between each vertex x in S_1 and its copy x' in S_1'. Then the transversal matroid on $S_1' \cup S_2$ corresponding to Γ' is isomorphic to the strict Menger matroid $M_{\Gamma, S_1}(S_1 \cup S_2)$. While transversal matroids are Menger matroids, not every Menger matroid is a transversal matroid. Indeed the matroid of rank 2 pictured affinely in Figure 5.3 is a strict Menger matroid (since it is the dual of a transversal matroid) but it is not a transversal matroid.

The class of Menger matroids being the class of restrictions of strict Menger matroids is clearly closed under restriction. On the other hand, by Theorem 5.3.2 the class of strict Menger matroids is identical to the class of duals of transversal matroids. Since the class of transversal matroids is closed under restriction, the class of duals of transversal matroids is closed under contraction. Hence the class of strict Menger matroids is closed under contraction. It now follows that the class of Menger matroids is closed under both restriction and contraction, that is, under taking minors. Thus the class of Menger matroids is a minor-closed class of matroids. It also follows now that the class of Menger matroids is identical to the class of contraction of transversal matroids.

We now obtain a proof of Menger's theorem for digraphs, which gives the rank function of a strict Menger matroid. The proof we give is due to Ingleton and Piff (1973) and is based on Lemma 5.3.1. If X, Y, and Z are sets of vertices of a digraph Γ, then Z *separates* X *from* Y provided every path with initial vertex in X and terminal vertex in Z has at least one of its vertices in Y.

5.3.3. Theorem. *(Menger's theorem). Let $\Gamma = \Gamma(S)$ be a digraph and let A and B be subsets of S. Then A can be linked into B if and only if no set of fewer than $|A|$ vertices separates A from B.*

Proof. We first note that if there is a linking Θ of A into B, then since Θ consists of $|A|$ pairwise vertex disjoint paths from A to B, a set of vertices which separates A from B has cardinality at least equal to $|A|$. It follows from Lemma 5.3.1 that A can be linked into B if and only if $S\backslash A$ contains a transversal of the family $\mathscr{A} = (A_s : s \in S\backslash B)$ where recall

$$A_s = \{t : st \text{ is an arc of } \Gamma\} \cup \{s\}.$$

By Hall's theorem, Proposition 4.1.4, $S\backslash A$ contains a transversal of $\mathscr{A}$ if and only if

$$\left|\left(\bigcup_{s\in X} A_s\right) \cap (S\backslash A)\right| \geqslant |X| \quad \text{for all} \quad X \subseteq S\backslash B. \tag{5.6}$$

Thus we need to show (5.6) is equivalent to the statement that no fewer than $|A|$ vertices separate A from B.

Suppose C separates A from B, and let A' be the set consisting of all those vertices which are separated from B by C. In particular $A \cup C \subseteq A'$. Let $X = A'\backslash C$. Then it follows that $X \subseteq S\backslash B$ and that $\cup_{s\in X} A_s \subseteq A'$. Hence

$$\left|\left(\bigcup_{s\in S} A_s\right) \cap (S\backslash A)\right| \leqslant |A' \cap (S\backslash A)| = |C| + |X| - |A|.$$

Hence if (5.6) holds, $|C| \geqslant |A|$. (In view of our earlier comment, this part of the proof is redundant.)

Now suppose (5.6) does not hold, so that there exists some $X \subseteq S \backslash B$ such that

$$\left| \left(\bigcup_{s \in X} A_s \right) \cap (S \backslash A) \right| < |X|.$$

Let $Z = \bigcup_{s \in X} A_s$ so that $X \subseteq Z$. Then $|Z| - |Z \cap A| < |X|$, so that

$$|Z| - |X| < |Z \cap A|.$$

Let $Y = Z \backslash X$. Then $|Y| < |Z \cap A|$, and since $X \subseteq S \backslash B$, Y separates Z from B and hence separates $Z \cap A$ from B. Therefore $C = Y \cup (A \backslash Z)$ separates A from B where

$$|C| \leqslant |Y| + |A \backslash Z| < |Z \cap A| + |A \backslash Z| = |A|.$$

Hence if (5.6) does not hold it is possible to separate A from B using fewer than $|A|$ vertices, and the theorem now follows. $\qquad \square$

5.3.4. Corollary. *If r denotes the rank function of the strict Menger matroid $M_{\Gamma,B}(S)$ then for each $A \subseteq S$, $r(A)$ is the minimum cardinality of a set of vertices which separates A from B.*

Proof. Let k be the minimum cardinality of a set of vertices which separates A from B. Let Γ' be the digraph obtained from Γ by adjoining a set A' of k new vertices with an arc from each of them to each of the vertices of A. Then k is also the minimum cardinality of a set of vertices which separates A' from B in Γ'. The result now follows readily from Theorem 5.3.3. $\qquad \square$

By Theorem 5.3.2 transversal matroids are precisely the duals of strict Menger matroids. Hence Proposition 5.2.9 can be used to give a characterization of strict Menger matroids. We first observe that the complements of the cyclic flats of a matroid are precisely the cyclic flats of its dual.

Let $M(S)$ be a matroid of rank k, and for a cyclic flat F, let $\mathscr{K}(F)$ denote the set of all cyclic flats which are properly contained in F. We define recursively an integer-valued function σ on the partially ordered set $\mathscr{F}$ of cyclic flats of $M(S)$ by $\sigma(\bar{\varnothing}) = 0$ and, for $\bar{\varnothing} \neq F \in \mathscr{F}$,

$$\sigma(F) = \left(|F| - r(F) - \sum_{K \in \mathscr{K}(F)} \sigma(D) \right)^+.$$

Let $\mathscr{B}^* = (F_1^*, \dots, F_m^*)$ be the family of cyclic flats in which each cyclic flat F of M occurs $\sigma(F)$ times in $\mathscr{B}^*$.

5.3.5. Proposition. *The matroid $M(S)$ is a strict Menger matroid if and only if*

$$r\left(\bigcup_{i \in I} F_i^* \right) \leqslant \left| \bigcup_{i \in I} F_i^* \right| - |I|, \quad \text{for all} \quad I \subseteq \{1, \dots, m\}.$$

Proof. By Theorem 5.3.2 $M(S)$ is a strict Menger matroid if and only if its dual $M^*(S)$ is a transversal matroid. Applying Proposition 5.2.9 to $M^*(S)$ we obtain the theorem. □

A different but related characterization of strict Menger matroids is due to Mason (1972), and other proofs of his theorem have been given by Ingleton and Piff (1973) and Kung (1978).

5.4. Other Properties and Generalizations

In this final section we discuss some additional properties of transversal matroids and their duals, and also mention without proof generalizations of some of the results in the preceding section.

Let $M(S)$ be a matroid and let B_1 and B_2 be bases. Then there always exist bijections $\sigma: B_1 \to B_2$ and $\tau: B_1 \to B_2$ such that $(B_1 \backslash \{x\}) \cup \{\sigma(x)\}$ and $(B_2 \backslash \{\tau(x)\} \cup \{x\}$ are bases for each $x \in B_1$. For $x \in B_1 \cap B_2$, necessarily $\sigma(x) = \tau(x) = x$. When $x \in B_1 \backslash B_2$, $\sigma(x) \in B_2 \backslash B_1$ and must be chosen so that x is an element of the unique circuit contained in $B_1 \cup \{\sigma(x)\}$. Similarly for $x \in B_1 \backslash B_2$, $\tau(x) \in B_2 \backslash B_1$ and $\tau(x)$ is in the unique circuit contained in $B_2 \cup \{x\}$. Given any $x \in B_1 \backslash B_2$ it is always possible to find a $y \in B_2 \backslash B_1$ such that both $(B_1 \backslash \{x\}) \cup \{y\}$ and $(B_2 \backslash \{y\} \cup \{x\}$ are bases, and it is natural to entertain the possibility that the bijections σ and τ above can be chosen so that $\sigma = \tau$. That this is not always possible can be seen by consideration of the cycle matroid of the complete graph on four vertices drawn in Figure 5.5 with edges labeled $1, 2, 3, 4, 5, 6$. For the bases $B_1 = \{1, 2, 3\}$ and $B_2 = \{4, 5, 6\}$, it is straightforward to check that σ and τ cannot be chosen to be equal. This leads to the following concept which was introduced in the work of Brualdi and Scrimger (1968) and Brualdi (1969). The matroid $M(S)$ is said to be *base orderable* if given any two bases B_1 and B_2 there is a bijection $\pi: B_1 \to B_2$ such that both $(B_1 \backslash \{x\}) \cup \{\pi(x)\}$ and $(B_2 \backslash \{\pi(x)\}) \cup \{x\}$ are bases for all $x \in B_1$. Such a bijection π is called a *base ordering bijection* for (B_1, B_2). Thus the cycle matroid of the complete graph K_4 is not base orderable but it is easy to show that the cycle matroid of every proper subgraph is.

The reason for our interest in base orderable matroids here comes from the following result of Brualdi and Scrimger (1968).

Figure 5.5. The complete graph K_4.

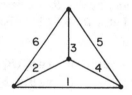

5.4.1. Lemma. *A transversal matroid is base orderable.*

Proof. Let $M(S)$ be the transversal matroid of rank k on S given by the bipartite graph $\Gamma(S, S_1)$. It follows from Lemma 5.1.1, using the correspondence between bipartite graphs and families of sets, that we may assume $|S_1| = k$. Let B_1 and B_2 be two bases of $M(S)$, and let Θ_1 and Θ_2 be matchings such that B_1 meets Θ_1 and B_2 meets Θ_2. Let $\Gamma' = \Gamma'(B_1 \cup B_2, S_1)$ be the bipartite graph with vertices as indicated whose edges are those in $\Theta_1 \cup \Theta_2$. Then each vertex $x \in B_1 \backslash B_2$ is in a connected component of Γ' which is a chain joining x to a vertex $\pi(x) \in B_2 \backslash B_1$. It is easy to see that this defines a bijection $\pi: B_1 \backslash B_2 \to B_2 \backslash B_1$ such that both $(B_1 \backslash \{x\}) \cup \{\pi(x)\}$ and $(B_2 \backslash \{\pi(x)\}) \cup \{x\}$ are bases for each $x \in B_1 \backslash B_2$. Extending π to B_1 by defining $\pi(x) = x$ for $x \in B_1 \cap B_2$, we conclude that $M(S)$ is base orderable. $\square$

By Lemma 5.4.1 the property of being base orderable is a necessary condition for a matroid to be a transversal matroid, but it is by no means sufficient. The cycle matroid $M(S)$ of the graph in Figure 5.6 is base orderable, but $M(S)$ is not a transversal matroid. The next two lemmas show that, unlike the class of transversal matroids, the class of base orderable matroids is a minor-closed class of matroids.

Figure 5.6. A graph whose cycle matroid is base orderable but not transversal.

5.4.2. Lemma. *Let $M(S)$ be a base orderable matroid. Then for each $T \subseteq S$, the restriction $M(T)$ is base orderable.*

Proof. Let A_1 and A_2 be bases of $M(T)$. Then there exists $X \subseteq S \backslash T$ such that both $B_1 = A_1 \cup X$ and $B_2 = A_2 \cup X$ are bases of $M(S)$. Let $\pi: B_1 \to B_2$ be a base ordering bijection for (B_1, B_2). Restricting π to A_1, we obtain a base ordering bijection for (A_1, A_2). We conclude $M(T)$ is base orderable. $\square$

5.4.3. Lemma. *The dual of a base orderable matroid is base orderable.*

Proof. Let $M(S)$ be a base orderable matroid, and let B_1' and B_2' be bases of its dual $M^*(S)$. Then $B_1 = S \backslash B_1'$ and $B_2 = S \backslash B_2'$ are bases of $M(S)$ where $B_1 \backslash B_2 = B_2' \backslash B_1'$ and $B_2 \backslash B_1 = B_1' \backslash B_2'$. Let $\pi: B_1 \to B_2$ be a base ordering bijection for (B_1, B_2). Define $\pi': B_2' \to B_1'$ by

$$\pi'(x) = \begin{cases} \pi(x) & \text{if } x \in B_2' \backslash B_1', \\ x & \text{if } x \in B_1' \cap B_2'. \end{cases}$$

Then π' is a base ordering bijection for (B_2', B_1'). It follows that $M^*(S)$ is base orderable. □

5.4.4. Corollary. *Each minor of a base orderable matroid is base orderable.*

Proof. This is an immediate consequence of Lemmas 5.4.2 and 5.4.3. □

5.4.5. Proposition. *A Menger matroid is a base orderable matroid.*

Proof. By Theorem 5.3.2, strict Menger matroids are duals of transversal matroids. Hence by Lemmas 5.4.1 and 5.4.3, a strict Menger matroid is base orderable. Since Menger matroids are restrictions of strict Menger matroids, the theorem now follows from Lemma 5.4.2. □

The class of Menger matroids and the class of base orderable matroids are both minor-closed classes of matroids with the former contained in the latter. This containment is proper. An example of a base orderable matroid that is not a Menger matroid can be obtained from the 9-point configuration which violates Pappus' theorem of projective geometry.

A property stronger than base orderability was shown to be true for transversal matroids by Brualdi and Scrimger (1968). A matroid $M(S)$ is said to be *strongly base orderable* if given any two bases B_1 and B_2 there exists a bijection $\pi : B_1 \to B_2$ such that both $(B_1 \backslash A) \cup \pi(A)$ and $(B_2 \backslash \pi(A)) \cup A$ are bases for all $A \subseteq B_1$. A base orderable matroid need not be strongly base orderable, although examples are not easy to find. The first example of a matroid which is base orderable but not strongly base orderable was found by Ingleton (1971).

5.4.6. Example. Let $M(S)$ be the matroid on $S = \{a_1, a_2, a_3, a_4, b_1, b_2, b_3, b_4\}$ whose bases are all 4-element subsets of S except for

$$\{a_1, b_1, b_2, b_4\}, \quad \{a_2, b_1, b_2, b_3\}, \quad \{a_1, a_3, a_4, b_3\}$$
$$\{a_2, a_3, a_4, b_4\}, \quad \{a_1, a_2, b_3, b_4\}.$$

It is not difficult to check that $M(S)$ is a matroid. It can be checked that $M(S)$ is base orderable. But $M(S)$ is not strongly base orderable since the defining property is not satisfied by the two bases $\{a_1, a_2, a_3, a_4\}$ and $\{b_1, b_2, b_3, b_4\}$.

Lemmas 5.4.1 to 5.4.3 and Corollary 5.4.4 remain true when base orderability is replaced by strong base orderability. Except for the obvious changes, the proofs are identical. In particular we conclude that Menger matroids are strongly base orderable.

In Chapter 7 of White (1986) the construction *matroid union* was defined and it was pointed out, indeed is a straightforward consequence of definitions, that a matroid is a transversal matroid if and only if it is a union of matroids of rank 1. In particular, since matroids of rank 1 are vector matroids over every field, it follows from Proposition 7.6.14 of White (1986) that transversal matroids are vector matroids over every sufficiently large field. Since strict Menger matroids are duals of transversal matroids and since the dual of a vector matroid over a field F is also a vector matroid over F, strict Menger matroids and hence Menger matroids are also vector matroids over every sufficiently large field.

Transversal matroids have matrix representations as vector matroids in which the non-zero entries are algebraically independent transcendentals over a field, say the real field $\mathbb{R}$. As a result, transversal matroids are vector matroids which are 'as free as possible'. To make this statement more precise, let $M(S)$ be a transversal matroid of rank k with presentation $\mathscr{A} = (A_1, \ldots, A_k)$. Let $|S| = n$ and let the elements of S be listed as $s_1, \ldots, s_n$. We then form the k by n incidence matrix $P = [p_{ij}]$ where for $i = 1, \ldots, k$ and $j = 1, \ldots, n$

$$p_{ij} = \begin{cases} 1 & \text{if } e_j \in S_i, \\ 0 & \text{if } e_j \notin S_i. \end{cases}$$

Let $X = [x_{ij}]$ be the matrix obtained from P by replacing the non-zero entries of P by algebraically independent transcendentals over R. We call X a *transcendental incidence matrix* of the family $\mathscr{A}$ of subsets of S. The following result was first observed by Edmonds (1967) and Mirsky and Perfect (1967).

5.4.7. Theorem. *Let $M(S)$ be a transversal matroid of rank k with presentation $\mathscr{A} = (A_1, \ldots, A_k)$, and let X be a transcendental incidence matrix of $\mathscr{A}$. Then A is an independent set of $M(S)$ if and only if the corresponding columns of X are linearly independent over $R(X)$, the field obtained by adjoining the transcendental entries of X to R.*

Proof. Let A be a subset of E consisting of the t elements $s_{i_1}, \ldots, s_{i_t}$. Let $X' = X[1, \ldots, k; i_1, \ldots, i_t]$ be the k by t submatrix of X corresponding to these elements. Then the columns of X' are linearly independent if and only if X' has a t by t submatrix with a non-zero determinant. Consider any t by t submatrix X'' of X', say the submatrix $X'' = X[1, \ldots, t; i_1, \ldots, i_t]$ formed by the first t rows of X'. Since the non-zero entries of X'' are algebraically independent transcendentals over R, $\det X'' \neq 0$ if and only if there is a permutation $j_1, \ldots, j_t$ of $1, \ldots, t$ such that $x_{j_1 i_1} \neq 0, \ldots, x_{j_t i_t} \neq 0$. The latter property is equivalent to the fact that $s_{i_1}, \ldots, s_{i_t}$ is a transversal of $(A_1, \ldots, A_t)$. It follows that X' has linearly independent columns if and only if A is a partial transversal of $\mathscr{A}$. $\square$

Suppose now that $M(S)$ is a principal transversal matroid of rank k with presentation $\mathscr{A} = (A_1, \ldots, A_k)$. Then the incidence matrix can be taken to have the form

$$P = [I_k | P']$$

where I_k is the k by k identity matrix. Let X be the corresponding transcendental incidence matrix so that the columns of X determine a vector matroid isomorphic to $M(S)$. (Actually the 1's of the identity matrix I_k above need not be replaced by transcendentals.) As shown by Brylawski (1975), $M(S)$ can be regarded as a special kind of affine matroid over R which is termed a *free simplicial affine matroid with spanning simplex B*. The set B of vertices of the simplex corresponds to the first k columns of X. Let c be any other column of X. Then c depends on a subset A of the first k columns of X, and we choose a point corresponding to c in the interior of the face $F(A)$ of the simplex determined by A. The points chosen on the faces corresponding to the columns of X are to be *freely situated* on the respective faces. This means the following. Let S' be the points of the simplex corresponding to the columns of X (the elements of S). Let p be a point in S' and suppose p is in the interior of the face $F(A)$ determined by the set A of vertices. Then p is *freely situated* on $F(A)$ if for all $Q \subseteq S'$ with $p \notin Q$, p is in the affine closure of Q if and only if $F(A)$ is in the affine closure of Q. For example, the affine matroid defined by Figure 5.7 is a free simplicial geometry with spanning simplex $B = \{b_1, b_2, b_3\}$. If in this picture p_2 were chosen so that p_2 was on the line joining p_1 and b_3, then p_2 would not be freely situated on the face whose interior contains it.

Figure 5.7. A free simplicial geometry.

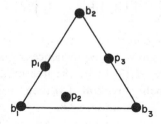

Since every transversal matroid is a restriction of a principal transversal matroid, it follows that a transversal matroid can be represented as a *free simplicial affine matroid* (some of the vertices of the simplex may be deleted). For more details on this construction, one may consult Brylawski (1975) or Brualdi and Dinolt (1975). In the latter paper a synthetic algorithm is given for obtaining a free simplicial affine matroid isomorphic to a given transversal matroid.

In the previous chapter we showed how a matroid induces a new matroid by means of a bipartite graph and we derived a formula for its rank function in terms of the rank function of the given matroid and the bipartite graph. This construction can be generalized by replacing the bipartite graph with an arbitrary digraph. The result is a generalization of (strict) Menger matroids and of Menger's theorem. We omit the proofs of the remaining results and refer the reader to the references given.

5.4.8. Proposition. *Let $\Gamma(S)$ be a digraph and let $M(S)$ be a matroid. Then the collection of subsets of S which are linked onto an independent set of $M(S)$ are the independent sets of a matroid $M'(S)$.*

This theorem was first proved by Brualdi (1971b) (see also Brualdi 1975) and Mason (1972). A proof using Lemma 5.3.1 was also given by Ingleton and Piff (1973). The rank function of the matroid $M'(S)$ above is a consequence of the following generalization of Menger's theorem due to Brualdi (1971a).

5.4.9. Proposition. *Let $\Gamma(S)$ be a digraph and let $M_1(S)$ and $M_2(S)$ be two matroids on S of equal rank with rank functions denoted by r_1 and r_2, respectively. Then the maximum k such that there are sets $A_1, A_2 \subseteq S$ of cardinality k such that A_1 is linked onto A_2, A_1 is an independent set of $M_1(S)$, and A_2 is an independent set of $M_2(S)$ equals*

$$min\{r_1(Z_1) + r_2(Z_2) + |Z_0|\}$$

where the minimum is taken over all triples (Z_0, Z_1, Z_2) such that $Z_1 \subseteq S, Z_2 \subseteq S$, and Z_0 separates $S \backslash Z_1$ and $S \backslash Z_2$ in $\Gamma(S)$.

As shown by McDiarmid (1972), the maximum evaluated in Proposition 5.4.9 also equals

$$\min_{Z \subseteq V}\left\{r_1\left(S \backslash \bigcup_{s \in Z} A_s\right) + \left|\left(\bigcup_{s \in Z} A_s\right) \backslash Z\right| + r_2(Z)\right\}$$

where as before, for $s \in S$, A_s consists of s and all those vertices t for which st is an arc. Welsh (1976, p. 226) shows the equivalence of the two expressions. If in Proposition 5.4.9 we take $M_1(S)$ to be the free matroid on S (whose rank function is then the cardinality function of S), then we obtain the rank function for the matroid $M'(S)$ of Proposition 5.4.8.

Finally we mention the following. Let G be a graph. Then Bondy (1972b) has shown that the cycle matroid of G is a transversal matroid if and only if G contains no subgraph homeomorphic from the complete graph K_4 on four vertices or the square C_k^2 of a cycle of length $k \geqslant 3$. (C_k^2 is the graph obtained from a cycle of length k by doubling each edge.) The cycle matroid of G is base orderable if and only if G contains no subgraph homeomorphic from K_4.

Notes

More information about the presentations of transversal matroids can be found in Bondy (1972a) and Brualdi and Dinolt (1972). A different approach to some of the properties of transversal matroids can be found in Dawson (preprint), who begins with Ingleton's characterization of transversal matroids given in Proposition 5.2.12. Mason (1972) was the first to consider what are called strict Menger matroids in this chapter. That these matroids have a fundamental role was an important discovery and his paper contains many insightful results. Piff and Welsh (1970) showed that a transversal matroid is a vector matroid over every sufficiently large field. Atkin (1972) gives a lower bound on the size in terms of the rank and number of elements. A generalization of some of the ideas in the characterization of transversal matroids given in Proposition 5.2.9 can be found in Brualdi (1974b).

Exercises

5.1. Prove that a transversal matroid of rank k has at most $\binom{k}{j}$ cyclic flats of rank j for $j = 0, 1, \ldots, k$. (Brualdi and Mason 1972)

5.2. Let $M(S)$ be a transversal matroid of rank k with presentation $(A_1, A_2, \ldots, A_k)$. Let the restriction $M(S \backslash A_1)$ have rank m and let $x \in A_1$. Suppose $x \in A_i (2 \leqslant i \leqslant t)$ and $x \notin A_i (t+1 \leqslant i \leqslant k)$. Prove that $(A_1 \backslash \{x\}, A_2, \ldots, A_k)$ is a presentation of $M(S)$ if and only if for some p with $2 \leqslant p \leqslant t$ the transversal matroid with presentation $(A_2 \backslash A_1, \ldots, A_{p-1} \backslash A_1, A_{p+1} \backslash A_1, \ldots, A_k \backslash A_1)$ has rank m. (Brualdi and Dinolt 1972)

5.3. Let $M(S)$ be a transversal matroid of rank k with maximal presentation $(M_1, \ldots, M_k)$. Suppose $(A_1, \ldots, A_k)$ is also a presentation of $M(S)$ where $A_i \subseteq M_i$ and $r(S \backslash A_i) = m_i$ for $i = 1, \ldots, k$. Prove that for each $i = 1, \ldots, k, |A_i|$ is the maximum cardinality of all subsets of M_i whose complement has rank m_i. (Brualdi and Dinolt 1972)

5.4. Let $M(S)$ be a transversal matroid with maximal presentation $(M_1, M_2, \ldots, M_k)$. Let C be a maximum cardinality cocircuit contained in M_1. Show by example that $(C, M_2, \ldots, M_k)$ need not be a presentation of $M(S)$.

5.5. Let $M(S)$ be a transversal matroid with maximal presentation $(M_1, M_2, \ldots, M_k)$. Let C_1 be a cocircuit of maximum cardinality contained in M_1. Prove that $(C_1, M_2, \ldots, M_k)$ is a presentation of $M(S)$ if and only if $(M_2 \backslash C_1, \ldots, M_k \backslash C_1)$ has a transversal. (Brualdi and Dinolt 1972)

5.6. Let $M(S)$ be a transversal matroid with maximal presentation $(M_1, \ldots, M_k)$. Let C_i be a cocircuit with $C_i \subseteq M_i$ for $i = 1, \ldots, k$. Prove that $(C_1, \ldots, C_k)$ is a (minimal) presentation of $M(S)$ if and only if for each $i = 1, \ldots, k$, C_i is a maximum cardinality cocircuit contained in M_i and $(C_1 \backslash C_i, \ldots, C_{i-1} \backslash C_i, C_{i+1} \backslash C_i, \ldots, C_k \backslash C_i)$ has a transversal. (Bondy 1972a)

5.7. Let $M(S)$ be a transversal matroid. Show by example that a cocircuit of $M(S)$ need not be a subset of some set of the maximal presentation.

5.8. Show by example that a truncation of a transversal matroid need not be a transversal matroid.

5.9. Prove that a truncation of a base orderable matroid is base orderable.

5.10. Prove that the dual of a principal transversal matroid is a principal transversal matroid. (Las Vergnas, see Brualdi 1974a)

5.11. Let $M(S)$ be a matroid where $|S| = n$. Prove that if $M(S)$ has rank at least $n - 2$, then $M(S)$ is a transversal matroid. Conclude that a matroid of rank 1 or 2 is the dual of a transversal matroid. (Ingleton and Piff 1973)

5.12. Prove that a Menger matroid of rank 3 is the dual of a transversal matroid.

5.13. Let $M(S)$ be a matroid. For a subset X of S let $\mathscr{F}(X)$ be the set of all flats properly contained in X. Define an integer-valued function α on subsets of S recursively by:

$$\alpha(X) = |X| - r(X) - \sum_{F \in \mathscr{F}(X)} \alpha(F).$$

Prove that $M(S)$ is a strict Menger matroid if and only if $\alpha(X) \geqslant 0$ for all $X \subseteq S$. (Mason 1972)

References

Atkin, A.O.L. (1972). Remark on a paper of Piff and Welsh. *J. Comb. Theory* **13**, 179–82.

Bondy, J.A. (1972a). Presentations of transversal matroids. *J. London Math. Soc.* (2), **5**, 289–92.

Bondy, J.A. (1972b). Transversal matroids, base orderable matroids, and graphs. *Quart. J. Math. (Oxford)* **23**, 81–9.

Bondy, J.A. and Welsh, D.J.A. (1971). Some results on transversal matroids and constructions for identically self dual matroids. *Quart. J. Math. (Oxford)* **22**, 435–51.

Brualdi, R.A. (1969). Comments on bases in dependence structures. *Bull. Australian Math. Soc.* **2**, 161–9.

Brualdi, R.A. (1971a). Menger's theorem and matroids. *J. London Math. Soc.* **4**, 46–50.

Brualdi, R.A. (1971b). Induced matroids. *Proc. Amer. Math. Soc.* **29**, 213–21.

Brualdi, R.A. (1974a). On fundamental transversal matroids. *Proc. Amer. Math. Soc.* **45**, 151–6.

Brualdi, R.A. (1974b). Weighted join-semilattices. *Trans. Amer. Math. Soc.* **191**, 317–28.

Brualdi, R.A. (1975). Matroids induced by directed graphs, a survey, *Recent Advances in Graph Theory*, Czech. Academy of Sciences, Prague, 115–34.

Brualdi, R.A. and Dinolt, G.W. (1972). Characterization of transversal matroids and their presentations. *J. Comb. Theory* **12**, 268–86.

Brualdi, R.A. and Dinolt, G.W. (1975). Truncations of principal geometries, *Discrete Math.* **12**, 113–38.

Brualdi, R.A. and Mason, J.H. (1972). Transversal matroids and Hall's theorem. *Pacific J. Math.* **41**, 601–13.

Brualdi, R.A. and Scrimger, E.B. (1968). Exchange systems, matchings, and transversals. *J. Comb. Theory* **5**, 244–57.

Brylawski, T.H. (1975). An affine representation for transversal geometries. *Studies in Applied Math.* **54**, 143–60.

Dawson, J. preprint.

Edmonds, J. (1967). Systems of distinct representatives and linear algebra. *J. Res. Nat. Bur. Stand.* **69B**, 241–5.

Ingleton, A.W. (1971). Conditions for representability and transversality of matroids. *Proc. Fr. Br. Conf. (1970)*, *Springer Lec. Notes*, **211**, 62–7.

Ingleton, A.W. (1975). Non base-orderable matroids. *Proc. Fifth British Combinatorial Conf.* (C. St. J.A. Nash-Williams and J. Sheehan, eds.), *Utilitas*, 355–9.

Ingleton, A.W. and Piff, M.J. (1973). Gammoids and transversal matroids. *J. Comb. Theory* **15**, 51–68.

Kung, J.P.S. (1978). The alpha function of a matroid–I. Transversal matroids. *Studies in Applied Math.* **58**, 263–275.

Mason, J.H. (1970). A characterization of transversal independence spaces. *Proc. Fr. Br. Conf. (1970)*, *Springer Lec. Notes*, **211**, 86–95.

Mason, J.H. (1972). On a class of matroids arising from paths in graphs. *Proc. London Math. Soc.* (3), **25**, 55–74.

McDiarmid, C. (1972). Strict gammoids and rank functions. *Bull. London Math. Soc.* **4**, 196–8.

Mirsky, L. and Perfect, H. (1967). Applications of the notion of independence to combinatorial analysis. *J. Comb. Theory* **2**, 327–57.

Perfect, H. (1968). Applications of Menger's graph theorem. *J. Math. Analysis Appl.* **22**, 96–111.

Piff, M.J. and Welsh, D.J.A. (1970). On the vector representation of matroids. *J. London Math. Soc.* **2**, 284–8.

Welsh, D.J.A. (1976). *Matroid Theory*, Academic Press, London.

White, N. (1986). *Theory of Matroids*. Cambridge University Press.

6

Simplicial Matroids

RAUL CORDOVIL AND BERNT LINDSTRÖM

6.1. Introduction

Given a finite set $A = \{a_1, a_2, \ldots, a_n\}$ and an integer k with $0 \leqslant k \leqslant n$, let $\binom{A}{k}$ denote the set of all k-element subsets of A. A k-element set will also be called a k-simplex, but we must warn the reader that topologists would prefer the name $(k-1)$-simplex since the topological realization has dimension $k-1$. Formal linear combinations of k-simplices in $\binom{A}{k}$ with coefficients from a field F give a vector space $F^{\binom{A}{k}}$ of dimension $\binom{n}{k}$ over F.

For $X \in \binom{A}{k}$ define the *boundary* $\partial X \in F^{\binom{A}{k-1}}$

$$\partial(\varnothing) = 0, \tag{6.1}$$

$$\partial(\{a_i\}) = 0, \tag{6.2}$$

$$\partial(\{a_{i_1}, \ldots, a_{i_k}\}) = \sum_{j=1}^{k} (-1)^{j-1} \{a_{i_1}, \ldots, \hat{a}_{i_j}, \ldots, a_{i_k}\}, \quad (i_1 < \cdots < i_k). \tag{6.3}$$

The roof $\hat{}$ over a letter means 'delete it'.

The boundary operation is extended by linearity to all elements of $F^{\binom{A}{k}}$:

$$\partial\left(\sum_{v=1}^{\bar{m}} c_v X_v\right) = \sum_{v=1}^{m} c_v \partial(X_v), \quad \text{where } c_1, \ldots, c_v \in F. \tag{6.4}$$

The following important property of the boundary operation is left as an easy exercise:

$$\partial^2 X = 0, \quad X \in F^{\binom{A}{k}}, \quad k \geqslant 1. \tag{6.5}$$

6.1.1. Definition. A subset $\{X_1, \ldots, X_m\} \subseteq \binom{A}{k}$ is independent in the *full simplicial matroid* $S_k^A[F]$ if $\partial(X_1), \ldots, \partial(X_m)$ are linearly independent over F. The restriction of $S_k^A[F]$ to a subset $E \subseteq \binom{A}{k}$ is a *k-simplicial geometry (matroid)* if $k \geqslant 2$ (if $k = 0$ or 1).

It is easy to prove that the matroids $S_k^A[F]$ and $S_k^B[F]$ are isomorphic when $|A| = |B| = n$. In particular, the linear order of A (used in (6.3)) does not matter. The matroid is therefore also denoted by $S_k^n[F]$, where n is called the *order*.

6.1.2. Example. Consider a finite simple graph with vertex set A and edge set $E \subseteq \binom{A}{2}$. The 2-simplicial geometry on E over a field F is the cycle matroid of the graph (which does not depend on F).

Sometimes it is desirable to order the elements of a simplex in a linear order different from the initial linear order of A. Consider a k-simplex $X = \{a_{i_1}, a_{i_2}, \ldots, a_{i_k}\}$, where $i_1 < i_2 < \cdots < i_k$, and assume that $a_{j_1}, a_{j_2}, \ldots, a_{j_k}$ is a permutation of X. Then we define the *oriented simplex*

$$(a_{j_1}, a_{j_2}, \ldots, a_{j_k}) = \text{sign} \begin{pmatrix} i_1 i_2 \cdots i_k \\ j_1 j_2 \cdots j_k \end{pmatrix} \{a_{i_1}, \ldots, a_{i_k}\}, \tag{6.6}$$

where the sign, $+$ or $-$, depends on the parity of the permutation. One can prove as an exercise

$$\partial(a_{j_1}, a_{j_2}, \ldots, a_{j_k}) = \text{sign} \begin{pmatrix} i_1 i_2 \cdots i_k \\ j_1 j_2 \cdots j_k \end{pmatrix} \partial(a_{i_1}, \ldots, a_{i_k}). \tag{6.7}$$

6.1.3. Example. Consider the triangulation of the real projective plane in Figure 6.1. It is easy to verify

$$\partial[(1, 2, 4) + (1, 2, 6) + (1, 4, 3) + (1, 5, 3) + (2, 3, 5) + (2, 3, 6)$$
$$+ (1, 6, 5) + (2, 5, 4) + (3, 4, 6) + (4, 5, 6)] = 2[(1, 2) + (2, 3) + (3, 1)] = 0$$

Figure 6.1. A triangulation of the real projective plane.

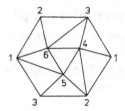

if and only if the characteristic of F is 2. The set of simplexes of the triangulation is therefore a circuit of $S_3^6[F]$ if and only if the characteristic of F is 2.

6.1.4. Example. If we triangulate a closed orientable surface, e.g. the topological 2-sphere in Figure 6.2, we will always get a circuit of a simplicial matroid, i.e., the characteristic of the field F does not matter.

Figure 6.2. A triangulated 2-sphere.

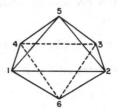

6.1.5. Proposition. *The rank of the simplicial matroid $S_k^n[F]$ is $\binom{n-1}{k-1}$. All circuits have cardinality at least $k+1$.*

Proof. For $k = 0, 1$, or n, the proposition is trivial. Then assume $2 \leqslant k \leqslant n-1$. Let $\mathscr{B}_a \subseteq \binom{A}{k}$ be the set of k-simplices containing a fixed $a \in A$. It is easy to see that $\{\partial(X): X \in \mathscr{B}_a\}$ is a linearly independent set of vectors. We claim that $\mathscr{B}_a$ is a base of $S_k^n[F]$.

Let $Z' \in \binom{A-a}{k}$. Then $\mathscr{B}_a \cup \{Z'\}$ contains $\binom{Z' \cup a}{k}$, which is the boundary of a k-simplex and therefore a circuit of $S_k^n[F]$ (cf. Example 6.1.4). It follows that $\mathscr{B}_a$ is a base of $S_k^n[F]$. Since $|\mathscr{B}_a| = \binom{n-1}{k-1}$, this is the rank.

Let C be any circuit of $S_k^n[F]$. Choose $Y \in C$. Then if $Z \in \binom{Y}{k-1}$ there is $X \in C, X \neq Y$ such that $Z \in \binom{X}{k-1}$. Distinct Z's give distinct X's. There are k Z's, hence at least k X's, $X \in C$. Therefore $|C| \geqslant k+1$.   □

6.2. Orthogonal Full Simplicial Matroids

The main result of this section was first proved in Crapo and Rota (1970) with the aid of Alexander's duality theorem for manifolds, applied to simplices. We give an elementary proof depending on matrix algebra. For another elementary proof see White (1986, Section 5.5.).

6.2.1. Theorem. *The orthogonal $S_k^n[F]^*$ of $S_k^n[F]$ is isomorphic to $S_{n-k}^n[F]$.*

The bijection $X \leftrightarrow A - X$ *with* $X \in \binom{A}{k}$ *and* $A - X \in \binom{A}{n-k}$ *gives the isomorphism.*

6.2.2. Definition. Let A be ordered by $a_1, a_2, \ldots, a_n$. Then we order k-simplices lexicographically. The *simplicial matrix* $S(n, k) = (s_{p,q})$ has rows and columns labeled by the sets of $\binom{A}{k-1}$ and $\binom{A}{k}$ respectively in the lexicographic order, with $s_{p,q} = 0$ if $p \not\subseteq q$ and $s_{p,q} = (-1)^j$ if $q - p = \{a_{i_j}\}$, $q = \{a_{i_0}, \ldots, a_{i_j}, \ldots, a_{i_{k-1}}\}$, $i_0 < i_1 < \cdots < i_{k-1}$, $k \geqslant 1$.

6.2.3. Example. It may be instructive to see a simple example. We choose $S(4, 3)$. Rows are labeled by 2-simplices, columns by 3-simplices in the lexicographic order.

$$
S(4, 3) =
\begin{array}{c}
 \{1,2,3\}\{1,2,4\}\{1,3,4\}\{2,3,4\} \\
\left[
\begin{array}{cccc}
1 & 1 & 0 & 0 \\
-1 & 0 & 1 & 0 \\
0 & -1 & -1 & 0 \\
1 & 0 & 0 & 1 \\
0 & 1 & 0 & -1 \\
0 & 0 & 1 & 1
\end{array}
\right]
\begin{array}{l}
\{1,2\} \\
\{1,3\} \\
\{1,4\} \\
\{2,3\} \\
\{2,4\} \\
\{3,4\}
\end{array}
\end{array}
$$

It follows easily by Definition 6.2.2 that $S(n, k)$ has the following block structure when $2 \leqslant k \leqslant n - 1$:

$$
S(n, k) = \left(\begin{array}{c|c}
-S(n-1, k-1) & 0 \\
\hline
I & S(n-1, k)
\end{array} \right). \tag{6.8}
$$

6.2.4. Definition. A matrix S with entries from a field F is a *coordinatization matrix* of a matroid $M(E)$ if the columns of S are labeled by the elements of E such that $\{e_1, \ldots, e_m\}$ is an independent set of $M(E)$ if and only if the column vectors labeled by $e_1, \ldots, e_m$ are linearly independent over F.

6.2.5. Proposition. *The matrix* $S(n, k)$ *with entries in a field* F *is a coordinatization matrix of the full simplicial matroid* $S_k^n[F]$.

Proof. This follows by (6.3) and the Definitions 6.1.1 and 6.2.2. $\qquad \square$

6.2.6. Proposition. *The matrix* $[I, S(n-1, k)]$ *with entries in* F *is a coordinatization matrix of the full simplicial matroid* $S_k^n[F]$.

Proof. The result follows by Proposition 6.2.5 and since the rank of the matrix $S(n, k)$ is $\binom{n-1}{k-1}$ by Proposition 6.1.5, which implies that the first $\binom{n-1}{k-2}$ rows of $S(n, k)$ are linear combinations of the $\binom{n-1}{k-1}$ last rows. $\qquad \square$

The transpose of a matrix S is denoted by S^t.

6.2.7. Proposition. *The matrix* $(-S(n-1,k)^t, I)$ *with entries in F is a co-ordinatization matrix of* $S_k^n[F]^*$, *and so is* $S(n,k+1)^t$.

Proof. The first statement follows from Proposition 1.3.1 and 6.2.6. The second statement follows by (6.8) (with $k+1$ in the place of k), since the last $\binom{n-1}{k+1}$ rows of the matrix $S(n,k+1)^t$ are linear combinations of the first $\binom{n-1}{k}$ rows since the rank of the matroid $S_{k+1}^n[F]$ is $\binom{n-1}{k}$. $\qquad\square$

6.2.8. Definition. Let S and S' be matrices with entries in the field F. We shall say that S and S' are *projectively equivalent*, and write $S \overset{P}{\sim} S'$ if S' can be obtained from S after a sequence of the following operations:

(a) add or delete a row of 0's,

(b) multiply the entries of a row or column by a non-zero element of F,

(c) add a scalar multiple of one row to another row.

This is equivalent to the definition of projective equivalence given in Section 1.2, except for operation (a), which is essentially trivial.

6.2.9. Definition. If two matrices S and S' over F and F' respectively coordinatize the same matroid, then we say that S and S' are *geometrically equivalent* and write $S \overset{G}{\sim} S'$.

It is clear that $S \overset{P}{\sim} S'$ implies $S \overset{G}{\sim} S'$. This implication is not in general reversible.

6.2.10. Definition. Given a matrix S let S^r denote the matrix which is obtained if we read the rows and columns of S in reverse order. The matrix S^r is called the *reverse* of S.

We now recall that $S(n,k) = (s_{p,q})$ has rows and columns labeled by $p \in \binom{A}{k-1}$ and $q \in \binom{A}{k}$, with $A = \{a_1,\ldots,a_n\}$. Define

$$\text{sign } B = \prod_{i:a_i \in B} (-1)^i \quad \text{when } B \subseteq A, \quad B \neq \varnothing,$$

$$\text{sign } \varnothing = 1.$$

Then we define the matrix $\bar{S}(n,k) = (\bar{s}_{p,q})$ by

$$\bar{s}_{p,q} = (\text{sign } p)(\text{sign } q)s_{p,q}.$$

We obviously have

$$\bar{S}(n,k)^t \overset{P}{\sim} S(n,k)^t. \tag{6.9}$$

By (6.8) we find easily for $2 \leqslant k \leqslant n - 1$

$$\overline{S}(n, k) = \left(\begin{array}{c|c} \overline{S}(n-1, k-1) & 0 \\ \hline -1 & -\overline{S}(n-1, k) \end{array} \right), \qquad (6.10)$$

$$S(n, k)^{rt} = \left(\begin{array}{c|c} S(n-1, k)^{rt} & 0 \\ \hline I & -S(n-1, k-1)^{rt} \end{array} \right). \qquad (6.11)$$

We shall prove

$$\overline{S}(n, k) = - S(n, n-k+1)^{rt}, \quad 1 \leqslant k \leqslant n. \qquad (6.12)$$

by induction on n for $n \geqslant 1$. The case $n = 1$ is trivial. Let $n > 1$. Assume that (6.12) holds for $n - 1$. If $k = 1$ or n then (6.12) holds by

$$\overline{S}(n, 1) = (-1, 1, -1, 1, \ldots,) = - S(n, n)^{rt},$$
$$\overline{S}(n, n) = (-1, -1, \ldots, -1)^t = - S(n, 1)^{rt}.$$

If $2 \leqslant k \leqslant n - 1$, then (6.12) follows by (6.10) and (6.11) using the induction hypothesis.

Proof of Theorem 6.2.1. By Proposition 6.2.7 we know that $S(n, k+1)^t$ is a coordinatization matrix of $S_k^n[F]^*$. By (6.9) and (6.12), we have

$$S(n, k+1)^t \overset{P}{\sim} \overline{S}(n, k+1)^t = - S(n, n-k)^r.$$

The last matrix represents $S_{n-k}^n[F]$ with the order of columns (and rows) reversed. The reversed lexicographic order of $\binom{A}{n-k}$ corresponds to the lexicographic order of the complements in $\binom{A}{k}$.   □

6.3. Binary and Unimodular Full Simplicial Geometries

The full simplicial geometries $S_2^n[F]$ and $S_{n-2}^n[F]$ are graphic and cographic respectively. It is well-known (see Section 2.6) that graphic and cographic matroids are unimodular (= regular), hence also binary. In this section we shall determine which full simplicial geometries $S_k^n[F]$, $2 \leqslant k \leqslant n - 2$, are (a) binary, (b) unimodular. The results are due to Cordovil and Las Vergnas (1979) and Lindström (1979).

6.3.1. Theorem. *$S_2^n[F]$ and $S_{n-2}^n[F]$ are binary matroids. $S_k^n[F]$, $3 \leqslant k \leqslant n - 3$, is binary if and only if the characteristic of F is 2. $S_k^n[F]$, $2 \leqslant k \leqslant n - 2$, is unimodular if and only if $k = 2$ or $n - 2$.*

The proof depends on three lemmata.

Table 6.1

	123	124	125	126	134	135	136	145	146	156
234	1	1			1					
235	1		1			1				
236	1			1			1			
245		1	α					1		
246		1		*					1	
256			1	*						1
345						1	*	*		
346						1	β		γ	
356							1	*		*
456								1	δ	*

6.3.2. Lemma. *The geometry $S_3^6[F]$ is binary only if the characteristic of F is 2.*

6.3.3. Lemma. *The geometry $S_3^6[F]$ is not unimodular.*

6.3.4. Lemma. *The geometry $S_3^6[F]$ is a minor of $S_k^n[F]$ when $3 \leqslant k \leqslant n - 3$.*

Proof of Lemma 6.3.2. Let $A = \{1, 2, 3, 4, 5, 6\}$. We consider $S_3^4[F]$. For brevity we will write ijk in place of $\{i, j, k\}$ and (ijk) in place of (i, j, k).

Let $C = \{123, 124, 135, 145, 235, 245\}$ and $C' = \{123, 126, 134, 145, 156, 236, 346, 456\}$. Both C and C' are triangulations of topological 2-spheres, hence circuits of the geometry $S_3^4[F]$. The reader may also verify that the symmetric difference $C \bigtriangleup C'$ is the triangulation of the real projective plane in Figure 6.1. The symmetric difference of two circuits of a binary matroid is either a circuit or a disjoint union of circuits of the matroid (Theorem 2.2.1). By Example 6.1.3 we conclude that the characteristic of F is 2.   □

Proof of Lemma 6.3.3. We recall that $S_3^4[F]$ has a base which consists of all 3-sets in $A = \{1, 2, 3, 4, 5, 6\}$ which contain some fixed element (say 1) (cf. proof of Proposition 6.1.5). Elements not in this base have fundamental circuits of size 4 with respect to the base. In Table 6.1 we show all non-zero entries of A^t, when (I, A) is a coordinatizing matrix.

If the matroid is unimodular, then the matrix may be chosen to be totally unimodular by Theorem 3.1.1, condition (6). We may assume that the first element in each row and each column is 1 (multiply all entries of the row or column by -1, if necessary).

The entries $\alpha, \beta, \gamma, \delta$ and also those indicated by $*$ are either 1 or -1. The 3×3 submatrices $\{234, 235, 245\} \times \{123, 124, 125\}$ and $\{234, 236, 346\} \times \{123, 134, 136\}$ have determinants $-1 - \alpha$ and $-1 - \beta$ respectively, which implies that $\alpha = -1$ and $\beta = -1$ by the total unimodularity. Then the determinants of the submatrices $\{235, 236, 245, 346, 456\} \times \{123, 125, 136, 145, 146\}$ and $\{234,

245, 346, 456} $\times$ {124, 134, 145, 146} are $-\gamma - \delta$ and $\gamma - \delta$ respectively. If $\gamma, \delta \in \{1, -1\}$ at least one of these determinants equals ± 2, which contradicts the total unimodularity. Therefore $S_3^6[F]$ can not be unimodular. $\square$

Proof of Lemma 6.3.4. Obviously $S_3^6[F]$ is a submatroid of $S_3^{6+m}[F]$, when $m \geqslant 0$. Then it follows by Theorem 6.2.1 that $S_3^6[F] \simeq S_3^6[F]^*$ is a minor of $S_{3+m}^{6+m}[F]$, hence also a minor of $S_{3+m}^n[F]$, when $n \geqslant 6 + m$, which was to be proved. $\square$

Proof of Theorem 6.3.1. If the characteristic of F is 2, then $S(n, k)$ gives a binary coordinatization of $S_k^n[F]$ by Proposition 6.2.5. $S_2^n[F]$ and $S_{n-2}^n[F]$ are graphic and cographic respectively, and therefore unimodular. The conditions are thus sufficient.

The necessity of the conditions follows by Lemmata 6.3.2–6.3.4 and since minors of binary (unimodular) matroids are binary (respectively unimodular).

          $\square$

Exercise 6.1 gives another proof that $S_3^6[Z_2]$ is not unimodular, since the Fano matroid and its orthogonal are not unimodular. In fact this shows that the restriction of $S_3^6[Z_2]$ to the 13 elements $C_1 \cup C_2 \cup C_3$ is not unimodular. Cordovil proved in his Ph.D. thesis that smaller submatroids of $S_3^6[Z_2]$ are unimodular.

6.4. Uniquely Coordinatizable Full Simplicial Matroids

The uniqueness results proved in this section were discovered by Cordovil (1978a, 1980). Uniquely coordinatizable matroids were studied by Brylawski and Lucas (1976).

6.4.1. Definition. A matroid M is *uniquely F-coordinatizable* if it can be coordinatized by a matrix over the field F, and all such coordinatizing matrices are projectively equivalent (cf. Definition 6.2.9).

The main results are

6.4.2. Proposition. $S_k^n[F]$ *is uniquely F-coordinatizable.*

6.4.3. Proposition. *If the matroid $S_k^n[F]$ is coordinatizable over a field F', then $S_k^n[F] = S_k^n[F']$.*

The bulk of the proofs of these results consists in the proof of the following proposition.

6.4.4. Proposition. *Let $S = S(n, k)$ be the simplicial matrix over F of characteristic $\neq 2$, and let T be a matrix over a field F'. Then (6.13) implies (6.14),* *where*

$$[I, S] \overset{G}{\sim} [I, T],\qquad\qquad\qquad (6.13)$$

$$S = D_1 T D_2,\qquad\qquad\qquad (6.14)$$

where D_1 and D_2 are non-singular diagonal matrices [we identify the matrix $S(n, k)$ over F and the matrix $S(n, k)$ over F'].

The following proposition will be useful [Proposition 2.2 of Brylawski and Lucas (1976)]. We may omit the proof.

6.4.5. Proposition. *Let* $[I, A]$ *and* $[I, A']$ *be matrices over the fields* F *and* F' *respectively. Then* $[I, A] \overset{G}{\sim} [I, A']$ *holds if and only if every subdeterminant of* A *vanishes exactly when the corresponding subdeterminant of* A' *vanishes. In particular, the entry* a_{ij} *of* A *is* 0 *if and only if the entry* a'_{ij} *of* A' *is* 0.

Proof of Proposition 6.4.4. If $k = 1$ or $n - 1$ the theorem is evident. Let $2 \leqslant k \leqslant n - 2$. There are non-singular diagonal matrices D_1 and D_2 over F' such that the first non-zero entries in each row and column of the matrix $R = D_1 T D_2$ are equal to the corresponding entries in the simplicial matrix $S(n, k)$. We shall prove that this implies the equality

$$R = S(n, k) = (s_{a,b}),\quad (a, b) \in \binom{A}{k-1} \times \binom{A}{k}.$$

We suppose that the non-zero entries of the matrix $R = (r_{a,b})$ are ordered by the lexicographic order of the indices (a, b). We prove by induction on this ordered set that $r_{a,b} = s_{a,b}$.

Suppose that $r_{a',b'} = s_{a',b'}$ holds when $(a', b') < (a, b)$. Let x be the first element of the set $A - b$, let $b - a = \{y\}$, and let z be the last element of the set b.

We have $x < y < z$. For if $y < x$ (respectively $y = z$) then $r_{a,b}$ is the first non-zero entry of the row a of the matrix $S(n, k)$ [respectively $r_{a,b}$ is the first non-zero entry of the column b of the matrix $S(n, k)$].

Let $a_1 = (a - \{z\}) \cup \{x\}$, $a_2 = (a - \{z\}) \cup \{y\}$, $b_1 = (b - \{z\}) \cup \{x\}$, $b_2 = (b - \{y\}) \cup \{x\}$. Let S_1 (respectively R_1) be the submatrix of S (respectively R) indexed by rows a_1, a_2, a and columns b_1, b_2, b. The non-zero entries $r_{a_1,b_1}, r_{a_1,b_2}, r_{a_2,b_1}, r_{a_2,b_2}, r_{a,b_2}$ of the matrix R are then equal to the corresponding elements of the matrix S by the hypothesis of induction because $a_1 < a_2 < a$ and $b_1 < b_2 < b$. If

$$b = \{\dots, \hat{x}, \dots, y, \dots, z\},$$
$$\phantom{b = \{\dots, }q-1k$$

$$b_1 = \{\dots, x, \dots, y, \dots\},\quad \text{and}\quad b_2 = \{\dots, x, \dots, \hat{y}, \dots, z\}$$
$$\phantom{b_1 = \{\dots, }pq, \text{and}\ b_2 = \{\dots, }pk$$

with elements in increasing order from left to right and position numbers indicated below, then the submatrix S_1 will be

$$S_1 = \begin{bmatrix} (-1)^{q+1} & (-1)^{k+1} & 0 \\ (-1)^{p+1} & 0 & (-1)^{k+1} \\ 0 & (-1)^{p+1} & (-1)^{q} \end{bmatrix} \begin{matrix} a_1 \\ a_2 \\ a \end{matrix} .$$

with column headings b_1, b_2, b.

Since $\det S_1 = 0$, the determinant of the corresponding submatrix of R_1 is also 0 by (6.13) and Proposition 6.4.5. Since $r_{a',b'} = s_{a',b'}$ when $(a', b') < (a, b)$, by the induction hypothesis, it follows that

$$r_{a,b} = -s_{a_1,b_1}s_{a_1,b_2}s_{a_2,b}s_{a,b_2} = (-1)^{q} = s_{a,b},$$

which was to be proved.   □

Proof of Theorem 6.4.2. If the characteristic of F is 2, then $S_k^n[F]$ is a binary matroid, and binary matroids are uniquely coordinatizable (by Proposition 6.4.5).

Then assume that the characteristic of F is not 2. We apply Proposition 6.4.4 with $F' = F$. Since (6.14) implies $[I, S] \overset{P}{\sim} [I, T]$ and $[I, S] \overset{G}{\sim} [I, T]$ implies (6.13), we conclude that $[I, S] \overset{G}{\sim} [I, T]$ and $[I, S] \overset{P}{\sim} [I, T]$ are equivalent. The theorem follows easily from this equivalence.   □

Proof of Theorem 6.4.3. If both F and F' have characteristic 2, then $S_k^n[F] = S_k^n[Z_2] = S_k^n[F']$.

If the characteristic of F is 2 and the characteristic of F' is distinct from 2, then the matroid is unimodular ($=$ regular) by a theorem of Tutte (cf. Brylawski 1975). It follows then by Theorem 6.3.1 that $S_k^n[F]$ is either graphic or cographic, and $S_k^n[F] = S_k^n[F']$ follows.

Finally, if the characteristic of F is not 2, the theorem follows by Proposition 6.2.6 and 6.4.4.   □

6.5. Matroids on the Bases of Matroids

It is well-known that the set $\mathscr{B}$ of bases of a matroid $M(E) = M$ of rank r is a subset of $\binom{E}{r}$. We may therefore consider the restriction of the full simplicial matroid, $S_r^E[F](\mathscr{B})$, which will be denoted by $S(M, F)$. This simplicial matroid was studied by Lindström (1981a). The main reason for studying this matroid was an interesting duality $S(G^*, F)^* \simeq H(G, F)^*$, where G is a geometry and $H(G, F)$ is a matroid on the bases of G, the definition of which depends on the order complex $\Delta(L)$ of the geometric lattice $L = L(G)$ associated with G.

Before we define $H(G, F)$, we shall consider $S(M, F)$ in some detail.

6.5.1. Definition. $S(M, F) = S_r^E[F](\mathscr{B})$, where $\mathscr{B}$ is the set of bases of $M = M(E)$, a matroid of rank r.

6.5.2. Proposition. $S(M^*, F)^* \simeq S_r^E[F]/\left(\binom{E}{r} - \mathscr{B}\right).$

Proof. Let $|E| = n$. Let $\mathscr{B}^*$ denote the set of bases of M^*. With the aid of Tutte's relation $(M/A)^* = M^* - A$ and Theorem 6.2.1, we get

$$\left(S_r^E[F]/\left(\binom{E}{r} - \mathscr{B}\right)\right)^* \simeq S_r^E[F]^*(\mathscr{B})$$

$$\simeq S_{n-r}^E[F](\mathscr{B}^*) = S(M^*, F). \qquad \square$$

6.5.3. Proposition. *The rank of the matroid* $S(M, F)$ *is* $|\mathscr{B}| - \tilde{\mu}(M^*)$, *where* $\mu(M^*)$ *is the Möbius invariant of the orthogonal* M^* *of* M $[\tilde{\mu}(M^*) = 0$ *if* M^* *has a loop,* $\tilde{\mu}(M^*) = |\mu(0, 1)|$ *in case of no loops, where* μ *is the Möbius function of the geometric lattice of* $M^*]$.

Proof (sketch). The independent sets of M form a simplicial complex $IN(M)$ of topological dimension $r - 1$. The Betti number $\beta_{r-1}(IN(M)) = \tilde{\mu}(M^*)$ by a theorem of Björner (see White 1988). By a rank formula for simplicial matroids of Crapo and Rota (1971) we have $r(\mathscr{B}) = |\mathscr{B}| - \beta_{r-1}(\mathscr{B})$ in $S_r^E[F]$. Hence $r(S(M, F)) = |\mathscr{B}| - \tilde{\mu}(M^*)$. $\qquad \square$

We consider now a finite geometric lattice L and its associated geometry G. The supremum operation in L will be denoted by $\vee$. The partial order in L is denoted by $<$. Let 0 and 1 denote the minimal and maximal element of L respectively. The linearly $(<)$-ordered subsets of $L - \{0, 1\}$ give the order complex $\triangle(L)$, the homology of which was first determined by Folkman (1966). Folkman proved that the Betti number $\beta_{r-2}(\triangle(L)) = \tilde{\mu}(G) = |\mu(0, 1)|$.

For sets $A = \{a_1, a_2, \ldots, a_r\}$ of atoms in L define

$$\beta(A) = \sum_\pi (-1)^{i(\pi)}(a_{\pi(1)}, a_{\pi(1)} \vee a_{\pi(2)}, \ldots, a_{\pi(1)} \vee \cdots \vee a_{\pi(r-1)}),$$

where the sum is over all permutations $\pi(1), \ldots, \pi(r)$ of $1, \ldots, r$ and $i(\pi)$ is the number of inversions of π. The terms of the sum are oriented simplices of size $r - 1$ of $\triangle(L)$ with coefficients in the fixed field F.

6.5.4. Definition. Let $H(G, F)$ be the vector matroid of all $\beta(B)$, where B is a base of G.

A direct computation verifies (Björner 1982, p. 117)

$$\partial \beta(A) = 0 \quad \text{for any } A = \{a_1, \ldots, a_r\} \text{ of atoms.}$$

Another computation verifies [Orlik and Solomon 1980, (3.8)]

$$\beta(A) = 0 \quad \text{when } A \text{ is dependent in } G.$$

Orlik and Solomon (1980, Theorem 4.3) proved that $\beta(B)$ with $B \in \mathscr{B}$

generate the homology group $H_{r-2}(\triangle(L))$ of rank $\tilde{\mu}(G)$. There is also a proof of this by Björner (1982, Theorem 4.2) using so-called neat base-families. Since $H_{r-2}(\triangle(L))$ is torsion-free, it follows (by the universal coefficient theorem) for any F

$$\text{rank } H(G, F) = \tilde{\mu}(G). \tag{6.15}$$

6.5.5. Proposition. *Let G be a geometry. Then for any field F*

$$H(G, F) \simeq S(G^*, F)^*.$$

Proof. When $A = \{a_1, \ldots, a_{r-1}\} \in \binom{E}{r-1}, E = E(G)$, define

$$\sigma(A) = \sum_{\pi} (-1)^{r-1+i(\pi)} (a_{\pi(1)}, a_{\pi(1)} \vee a_{\pi(2)}, \ldots, a_{\pi(1)} \vee \cdots \vee a_{\pi(r-1)}),$$

where the sum is over all permutations π of $1, \ldots, r-1$. The following identity can easily be verified:

$$\sigma(\partial(B)) = \beta(B), \quad B \in \binom{E}{r}. \tag{6.16}$$

Let V_1, V_2, V_3 be vector spaces over F generated respectively by $\{\partial(B): B \in \mathscr{B}\}, \{\partial(B): B \in \binom{E}{r} - \mathscr{B}\}$, and $\{\beta(B): B \in \mathscr{B}\}$, where $\mathscr{B}$ is the set of bases of G.

The map σ can be extended to a linear map σ of V_1 onto V_3, by equation (6.16). Since $\sigma(V_2) = 0$, there is an induced linear map $\bar{\sigma}: V_1/(V_1 \cap V_2) \to V_3$ (onto). Therefore there is a strong map of the matroid $S_r^E[F]/\left(\binom{E}{r} - \mathscr{B}\right)$ onto $H(M, F)$. The rank of the first mentioned matroid is $\tilde{\mu}(G)$ by Propositions 6.5.2 and 6.5.3, which is also the rank of $H(G, F)$ by equation (6.15). The strong map is therefore an isomorphism. The theorem then follows by Proposition 6.5.2. $\qquad\square$

We may mention that A. Björner (1982, Proposition 6) has proved that the geometry $H(G, F)$ is 2-partitionable. A geometry $G(E)$ is said to be *2-partitionable* if for every $x \in E$ there is a partition $E - \{x\} = E_1 \cup E_2, E_1 \cap E_2 = \varnothing$, such that $x \notin \bar{E}_1$ and $x \notin \bar{E}_2$ ($\bar{E}_i$ is the span of E_i in G). M. Feinberg (1972) proved that a 2-partitionable vector geometry of rank r can not contain more than $\binom{r+1}{2}$ elements. It follows then by equation (6.15) that any geometry G has at most $\binom{\mu+1}{2}$ bases, where $\mu = \tilde{\mu}(G)$ is the Möbius invariant of G.

6.5.6. Example. We show by an example that the matroid $H(G, F)$ is not necessarily unimodular. Let $G = U_3^6$ be the uniform geometry of rank 3 on 6 elements. Then $(U_3^6)^* = U_3^6$. Note that $S(U_3^6, F) = S_3^6[F]$, and $H(U_3^6, F) \simeq S_3^6[F]$ by Proposition 6.5.5. Then see Lemma 6.3.3!

6.6. Sperner's Lemma for Geometries

We recall the classical lemma in topology discovered by E. Sperner.

Sperner's lemma. Let $\triangle$ be a triangulation of a topological d-sphere. Label the vertices of $\triangle$ by elements in a set E of size $d + 1$. Then if some d-dimensional simplex of $\triangle$ is labeled by all elements of E, then there are at least two simplices of $\triangle$ labeled by all elements of E.

Lovász (1980) observed that if we label the vertices of $\triangle$ by elements of a geometry G of rank $d + 1$ such that at least one simplex is labeled by all elements of base of G, then at least two simplices of $\triangle$ are labeled in this way.

This generalization of Sperner's lemma was extended to cycles of simplices over a field F by Lindström (1981b). Cordovil (1982b) observed that one can not use loops as labels, i.e., the result does not hold for matroids in general!

A set of r-simplices $\sigma_1, \ldots, \sigma_n$ is called an r-*cycle* over F, if for non-zero $\alpha_i \in F$, $\sum_{i=1}^n \alpha_i \partial(\sigma_i) = 0$.

6.6.1. Proposition. *If the points of an r-cycle are labeled by elements of a geometry G of rank r and some simplex is labeled by all elements of a base of G, then at least two simplices are labeled by entire bases of G.*

Proof. Let $f: V \to E$ be the labeling of vertices by elements of the geometry $G(E)$. If $(v_1, \ldots, v_r) = \sigma$ is a simplex, let $f(\sigma) = (f(v_1), \ldots, f(v_r))$ and extend the map to linear combinations of simplices. Note that the operators f and ∂ commute. We have now $\sum_{i=1}^n \alpha_i \partial(f(\sigma_i)) = 0$. Apply the linear operator σ of Proposition 6.5.5 (proof) and (6.16) gives $\sum_{i=1}^n \alpha_i \beta(f(\sigma_i)) = 0$. If $f(\sigma_j)$ is a base of G, then $\beta(f(\sigma_j)) \neq 0$. Then for some $k \neq j$, $\beta(f(\sigma_k)) \neq 0$, which implies that $f(\sigma_k)$ is also a base of G.

Kryński (1983) observed that Lovász' generalization of Sperner's lemma follows easily from this lemma and showed that Proposition 6.6.1 is a consequence of a generalized Sperner lemma by Sperner (1980). Cordovil (1982b) has an elementary proof of Proposition 6.6.1, which depends on an equivalence with the following interesting result.

6.6.2. Proposition. *Let $M(E)$ be a matroid of rank r without isthmus. Then, for every field F, the simplicial matroid $S(M, F) = S_r^E[F] (\mathscr{B})$ on the set $\mathscr{B}$ of bases of M is also a matroid without isthmus.*

One can even show that $S(M, F)$ is an inseparable matroid using the method of Cordovil (1982b).

6.7. Other Results

Crapo and Rota (1970) considered simplicial matroids over the rational numbers. Simplicial matroids over prime fields were studied by R. Reid (1970), but his results were not published. For the following theorem of Reid there is an elementary proof by Cordovil (1982a).

6.7.1. Proposition. *Let M be a matroid representable over a prime field F. Then there is a 3-simplicial matroid M′ over F, which is a series extension of M.*

Binary simplicial matroids were studied by M.J. Todd (1976b). There is also a paragraph in the book by Welsh (1976) on simplicial matroids over Z_2. See also Todd (1976a) and Tüma (1984).

E.D. Bolker (1976) applied simplicial matroids over the rational numbers in studying transportation polytopes.

Cordovil (1978a) proved the formula for rank in k-simplicial matroids and any F (β_k is the k:th Betti number):

$$r(X) = |X| - \beta_k(X) = \binom{n-1}{k-1} - \beta_{k-1}(X), \quad X \subseteq \binom{A}{k}.$$

In Crapo and Rota (1970) this is the definition of $r(X)$ and it is proved that this gives a rank function of a geometry on $\binom{A}{k}$.

Exercises

6.1. Verify that the following three circuits of $S_3^6[Z_2]$ generate a Fano matroid F_7 (see Figure 1.9):

$$C_1 = \{123, 124, 134, 234\},$$

$$C_2 = \{123, 124, 135, 145, 235, 245\},$$

$$C_3 = \{123, 126, 134, 145, 156, 236, 346, 456\}.$$

Also verify that the minor

$$(S_3^6[Z_2]/\{156, 235, 236, 245, 346, 456\})$$

$$(\{123, 124, 126, 134, 135, 145, 234\})$$

is isomorphic to the orthogonal of the Fano matroid F_7^*.

6.2. Given a finite simple graph $G = (V, E)$, let p and q be two points not in V. Consider the set S of all 2-simplices $\{u, v, p\}, \{u, v, q\}$ for $\{u, v\} \in E$. Prove that the 2-simplicial matroid over F on S is isomorphic to a graphic matroid.

6.3. From the simplicial matroid $S_3^6[F]$ delete four simplices which contain two of the points (e.g. the simplices $\{1, 2, 3\}, \{1, 2, 4\}, \{1, 2, 5\}, \{1, 2, 6\}$). Prove that the restriction of the matroid to the 16 remaining simplices is isomorphic to a cographic matroid and therefore regular.

6.4. (For those who know homology.) Assume that the bases of a simplicial matroid have torsion-free homology. Prove that the matroid is regular.

6.5. The triangulation of the real projective plane in Figure 6.1 contains 10 of the simplices of $S_3^6[Z_2]$. Verify that the remaining simplices give another triangulation of the real projective plane and show that these triangulations give two circuits which are also cocircuits of the matroid $S_3^6[Z_2]$.

6.6. Let $N_1, N_2, \ldots, N_k$ be disjoint points sets with $N_i = n_i$ $(1 \leqslant i \leqslant k)$. The direct product $N_1 \times N_2 \times \ldots \times N_k$ contains $n_1 n_2 \cdots n_k$ k-simplices. Consider the k-simplicial matroid of these k-simplices over a field F. Prove that the rank of the matroid is $n_1 n_2 \cdots n_k - (n_1 - 1)(n_2 - 1) \cdots (n_k - 1)$. Prove that the matroid is regular when $n_1 = n_2 = \cdots n_k = 2$. These matroids with $F = \mathbb{R}$ (the real numbers) occur implicitly in Bolker (1976). One can prove that the 3-simplicial matroid with $n_1 = n_2 = n_3 = 3$ is non-regular. (Lindström 1983)

6.7. Even if G is unimodular, $H(G, Z_2)$ need not be unimodular. Let $G = M(K_4)$, the cycle matroid of the complete graph K_4. The edges of K_4 can be labeled such that $C_1 \cup C_2 \cup C_3$ gives 13 spanning trees in K_4, where C_1, C_2, C_3 generate a Fano configuration (see Exercise 6.1). (Lindström 1983)

References

Björner, A. (1982). On the homology of geometric lattices. *Algebra Universalis* 14, 107–28.

Bolker, E.D. (1976). Simplicial geometry and transportation polytopes. *Trans. Amer. Math. Soc.* 217, 121–42.

Brylawski, T.H. (1975). A note on Tutte's unimodular representation theorem. *Proc. Amer. Math. Soc.* 52, 499–502.

Brylawski, T.H. and Lucas, D. (1976). Uniquely representable combinatorial geometries, in *Teorie Combinatorie Tomo I* (B. Segre, ed.), pp. 83–104. Accademia Nazionale dei Lincei, Rome.

Cordovil, R. (1978a). Sur les géométries simpliciales. *C.R. Acad. Sci. Paris Sér. A* 286, 1219–22.

Cordovil, R. (1978b). Matrices simpliciales en tant que généralisation des matrices totalement unimodulaires. *Actes du Colloq. Journées d'Algebre Appliquée et Combinatoire*, pp. 133–41. Grenoble.

Cordovil, R. (1980). Representation over a field of full simplicial matroids. *Europ. J. Combinatorics* 1, 201–5.

Cordovil, R. (1982a). On Reid's 3-simplicial matroid theorem. *Combinatorica* 2(2), 135–41.

Cordovil, R. (1982b). On simplicial matroids and Sperner's lemma, in *Proc. Colloquium on Matroid Theory*, 40, *Matroid Theory*, Szeged, Hungary, 97–105.

Cordovil, R. and Las Vergnas, M. (1979). Géométries simpliciales unimodulaires. *Discrete Math.* 26, 213–17.

Crapo, H.H. and Rota, G.-C. (1970). *On the Foundations of Combinatorial Theory: Combinatorial Geometries* (preliminary edition), MIT Press, Cambridge, Mass.

Crapo, H.H. and Rota, G.-C. (1971). Simplicial geometries. *Proc. Symp. Pure Math. XIX, Combinatorics.* Amer. Math. Soc. Publ.

Feinberg, M. (1972). On chemical kinetics of a certain class. *Arch. Rat. Mech. Anal.* 46, 1–41.

Folkman, J. (1966). The homology groups of a lattice. *J. Math. Mech.* 15(4), 631–6.

Kryński, S. (1983). Remarks on matroids and Sperner's lemma. *Polish Acad. Sci., preprint* MDP–1/83.

Lindström, B. (1979). Non-regular simplicial matroids. *Discrete Math.* **28**, 101–2.

Lindström, B. (1981a). Matroids on the bases of simple matroids. *Europ. J. Combinatorics* **2**, 61–3.

Lindström, B. (1981b). On matroids and Sperner's lemma. *ibid.* 65–6.

Lindström, B. (1983). Some examples of non-regular simplicial matroids. *Ars Combinatoria* **16A**, 199–204.

Lovász, L. (1980). Matroids and Sperner's lemma. *Europ. J. Combinatorics* **1**, 65–6.

Orlik, P. and Solomon, L. (1980). Combinatorics and topology of complements of hyperplanes. *Invent. Math.* **56**, 167–89.

Reid, R. (1970). Polyhedral and simplicial geometries (unpublished).

Sperner, E. (1980). Fifty years of further development of a combinatorial lemma. Part A and B, in *Numerical Solution of Highly Nonlinear Problems* (W. Forster, ed.), pp. 183–217. North-Holland.

Todd, M.J. (1976a). A combinatorial generalization of polytopes. *J. Comb. Theory Ser. B* **20**, 229–42.

Todd, M.J. (1976b). Characterizing binary simplicial matroids. *Discrete Math.* **16**, 61–70.

Tüma, J. (1984). Note on binary simplicial matroids. *Discrete Math.* **49**, 105–7.

Tutte, W.T. (1965). Lectures on matroids. *J. Res. Nat. Bur. Stand.* **69B**, 1–47.

Welsh, D.J.A. (1976). *Matroid Theory.* Academic Press, London.

White, N. (1986). *Theory of Matroids.* Cambridge University Press.

White, N. (1988). *Combinatorial Geometries: Advanced Theory.* Cambridge University Press, to appear.

7

The Möbius Function and the Characteristic Polynomial

THOMAS ZASLAVSKY

The effort to generalize graph theory to matroids has yielded analogs of the chromatic polynomial and related graph invariants and (although there is still no exact analog for an arbitrary matroid) a partial extension of vertex coloring. The 'characteristic polynomial' provides every matroid with an algebraic analog of the chromatic polynomial; Crapo and Rota's 'critical problem' defines a kind of proper coloring for submatroids of finite vector spaces. We shall begin our account with the characteristic polynomial, its logical building block the combinatorial Möbius function, and the related beta invariant; then we present examples including the connection with graph coloring and conclude with the critical problem.

As usual in enumeration we assume throughout this chapter that all matroids, lattices, and other combinatorial objects are finite.

7.1. The Möbius Function

The combinatorial Möbius function, which we will need for geometric lattices, can just as easily be defined for any finite partially ordered set. Let P be such a set and consider integral functions $P \times P \to \mathbb{Z}$. The function μ (or μ_P) which satisfies

$$\sum_{x \leqslant y \leqslant z} \mu(x, y) = \delta(x, z) \quad \text{if} \quad x \leqslant z \tag{7.1}$$

(where δ is the Kronecker delta) together with ordering property

$$\mu(x, z) = 0 \quad \text{if} \quad x \nleqslant z$$

is called the *Möbius function* of P. [Hall (1936). Weisner (1935) for lattices. The basic reference is Rota (1964). A good recent treatment is Aigner (1979).]

To see that μ exists and is uniquely defined, let us rewrite (7.1) as two equations:

$$\mu(x, x) = 1, \tag{7.2}$$

$$\mu(x, z) = - \sum_{x \leqslant y < z} \mu(x, y) \quad \text{if} \quad x < z. \tag{7.3}$$

We can calculate $\mu(x, z)$ first for $z = x$ from (7.2), then recursively from (7.3) for successively higher z by induction on the length of the longest chain from x to z. Thus the value of $\mu_P(x, z)$ depends only on the order structure of the interval $[x, z]$ and not on the rest of P.

To understand the Möbius function better, let us introduce the incidence algebra $I(P)$: the set of all functions $\phi : P \times P \to \mathbb{Z}$ such that $\phi(x, y) = 0$ if $x \not\leqslant y$, with pointwise addition and the convolution product

$$(\phi * \psi)(x, z) = \sum_{x \leqslant y \leqslant z} \phi(x, y) \psi(y, z).$$

This product is a form of matrix multiplication. If we extend $\leqslant_P$ to a linear ordering of P denoted by subscripts, so $p_i <_P p_j$ implies $i < j$, then an incidence function is a $|P|$ by $|P|$ upper-triangular matrix and convolution is matrix multiplication. Hence multiplication is associative and has δ for identity. Also, any incidence function ϕ with $\phi(x, x) \equiv 1$ is invertible. The *zeta function* of P is the function $\zeta \in I(P)$ with

$$\zeta(x, y) = 1 \quad \text{if} \quad x \leqslant y.$$

We can now restate the definition (7.1): *μ is the left inverse of ζ.*

The recursive formula (7.3) is an effective way to compute the Möbius function of a small interval. Some useful values are the following.

7.1.1. Proposition. *In a partially ordered set,*

$\mu(x, x) = 1,$

$\mu(x, y) = -1 \quad$ *if y covers x,*

$\mu(x, z) = n - 1 \quad$ *if $[x, z]$ is an n-point line.*

Proof. Exercise. $\qquad\qquad\qquad\qquad\qquad\qquad\qquad\qquad\qquad\qquad\square$

Before we concentrate on geometric lattices, we shall give some important general properties of the Möbius function.

7.1.2. Proposition. *The Möbius function of P can be defined by replacing (7.1) by*

$$\sum_{x \leqslant y \leqslant z} \mu(y, z) = \delta(x, z) \quad \text{if} \quad x \leqslant z,$$

116   Thomas Zaslavsky

or by replacing (7.3) *by*

$$\mu(x,z) = - \sum_{x<y\leqslant z} \mu(y,z) \quad if \quad x < z.$$

Proof. Exercise.   □

The *raison d'être* of the Möbius function is the inversion property. This is the common generalization of the principle of inclusion and exclusion (which is Möbius inversion on the power set of a set) and of number-theoretic Möbius inversion [in which P is the set of natural numbers ordered by divisibility; the classical $\mu(n) = \mu_P(1,n)$]. If $\phi \in I(L)$ and $f : P \rightarrow A$ (an abelian group, which will often be the integers), define

$$(\phi * f)(x) = \sum_{y \geqslant x} \phi(x,y)f(y),$$

$$(f * \phi)(y) = \sum_{x \leqslant y} f(x)\phi(x,y).$$

These are the incidence-algebra versions of the product of a vector by a matrix.

7.1.3. Proposition (Möbius Inversion). *Let P be a finite poset. Let f and g be functions on P with values in any ring (or $\mathbb{Z}$-module, i.e., abelian group). Then*

$$g(x) = \sum_{y \geqslant x} f(y)$$

implies

$$f(x) = \sum_{y \geqslant x} \mu_P(x,y)g(y),$$

and vice-versa. In addition

$$g(y) = \sum_{x \leqslant y} f(x)$$

implies

$$f(y) = \sum_{x \leqslant y} g(x)\mu_P(x,y),$$

and vice versa.

Proof. Exercise.   □

Now we specialize to the case of a finite matroid $M = M(E)$. Its lattice of flats L has Möbius function μ_L. [The value $\mu_L(0,1)$ is often called the *Möbius invariant* of L and written $\mu(L)$. As we noted earlier, the Möbius invariant of an interval $[x,y]$ in L, $\mu([x,y])$, is equal to $\mu_L(x,y)$.] The *Möbius function of M* is defined by

$$\mu_M(X,F) = \mu_L(X,F) \quad if \quad X, F \in L,$$
$$\mu_M(X,F) = 0 \qquad\quad if \quad X \notin L, \quad F \in L;$$

$\mu_M(X,F)$ is not defined if $F \notin L$. The purpose of this extended definition of μ_M is to allow matroids in which $\varnothing$ is not closed to obey the same formulas as other matroids–the same reason for which the chromatic polynomial of a graph with loops is taken to be identically 0. (These two cases are virtually the same, as our discussion of the chromatic polynomial will show.)

One such formula of basic importance is the following expansion (valid more generally for any closure on S). It seems to originate with Weisner. The non-trivial case (where $W \in L$) is a special case of Weisner (1935), Equation (15) with property $P' = $ 'minimal'. For a graphic matroid it is implied by a result of Whitney (1932).

7.1.4. Proposition (Boolean Expansion Formula). *Let L be the lattice of flats of the matroid $M = M(E)$. Let $W \subseteq E$ and $F \in L$. Then*

$$\mu_M(W, F) = \sum_{\substack{W \subseteq X \subseteq F \\ \mathrm{cl}\, X = F}} (-1)^{|X - W|}.$$

Proof. See Exercise 7.9. $\qquad\qquad\qquad\qquad\qquad\qquad\qquad\qquad\square$

7.1.5. Example. *Uniform matroids, Boolean algebras, and circuits.* In the uniform matroid U_{rm} of rank r on an m-set E, the flats of rank k (for $k < r$) are the k-subsets of E. We have

$$\mu(U_{rm}) = \sum_{k=0}^{r-1} (-1)^{k+1} \binom{m}{k}, \quad \text{if} \quad 0 < r \leqslant m.$$

In particular for the Boolean algebra $B_m = U_{mm}$ we have $\mu(B_m) = (-1)^m$. For the m-point circuit $C_m = U_{m-1, m}$ we have $\mu(C_m) = (-1)^{m-1}(m-1)$. (Exercise.)

Another useful formula, also valid for any closure, is:

7.1.6. Proposition. [*Special case of Weisner's theorem (Weisner 1935, Theorem 9; Rota 1964, p. 351, Corollary)*] *In the matroid $M = M(E)$ let F be a flat, e a point in F, and $F_1, F_2, \ldots$ the flats such that F covers F_i and $e \notin F_i$. Then*

$$\mu_M(\varnothing, F) = -\sum_i \mu_M(\varnothing, F_i).$$

Proof. For fixed e and any flat F containing e, let

$$f(F) = \mu_M(\varnothing, F) + \sum_i \mu_M(\varnothing, F_i).$$

We want to show $f \equiv 0$. Since that is trivially true if $\varnothing$ is not closed, we may assume $\mathrm{cl}\,(e)$ is an atom A in the lattice L of flats. Let

$$g(F) = \sum_{A \leqslant F' \leqslant F} f(F')$$

$$= \sum_{F'} \left[\mu(\varnothing, F') + \sum_i \mu(\varnothing, F'_i) \right].$$

Each flat $E' \leqslant F$ appears exactly once in the latter sum. For if $E' \geqslant A$, then E' is an F'; but if $E' \not\geqslant A$, then E' is one of the F'_i associated with that F' which equals $E' \vee A$. Hence

$$g(F) = \sum_{E' \leqslant F} \mu(\varnothing, E') = 0$$

since $F \geqslant A > 0$. We therefore have

$$\sum_{F' \in [A,F]} f(F') = g(F) = 0.$$

Applying Möbius inversion (Proposition 7.1.3), we see that $f \equiv 0$.   □

Now we are ready to prove the main properties of the Möbius function of a matroid. The first theorem is the core of Brylawski (1972, Theorem 4.2), which will reappear in Theorem 7.2.4 and Section 7.4.

7.1.7. Theorem. *The Möbius invariant of a matroid $M = M(E)$ satisfies:*
(i) *the deletion-contraction rule: if $e \in E$ is not an isthmus,*

$$\mu(M) = \mu(M - e) - \mu(M/e);$$

(ii) *the direct sum rule: if $M = M_1 \oplus M_2$, then*

$$\mu(M) = \mu(M_1)\mu(M_2).$$

Proof of (i). If e is a loop, $M - e = M/e$ and both sides of the equation are 0. Suppose then that e is neither an isthmus nor a loop.

We rewrite the left-hand side by Proposition 7.1.4:

$$\mu(M) = \sum_{\substack{X \subseteq E \\ \mathrm{cl} X = E}} (-1)^{|X|}$$

$$= \sum_{\substack{X \subseteq E - e \\ \mathrm{cl} X = E}} (-1)^{|X|} - \sum_{\substack{e \in X \subseteq E \\ \mathrm{cl} X = E}} (-1)^{|X - e|}. \qquad (7.4)$$

Since e is not an isthmus, a set $X \subseteq E - e$ spans M if and only if it spans $M - e$. Hence the first sum in (7.4) equals $\mu(M - e)$. The second sum equals $\mu(M/e)$, for an X containing e spans M if and only if $X - e$ spans M/e.

Proof of (ii). See Exercise 7.10.   □

Next is the fundamental theorem on the sign of μ in a geometric lattice.

7.1.8. Theorem. (*Rota 1964, Theorem 4, p. 357*) *The Möbius function of a geometric lattice L is non-zero and alternates in sign. Precisely,*

$$(-1)^{r(y)-r(x)}\mu_L(x, y) > 0 \quad \text{if} \quad x \leqslant y \quad \text{in} \quad L.$$

Proof. It will suffice to prove

$$(-1)^{r(L)}\mu_L(0,1) > 0. \qquad (7.5)$$

We use induction on the rank and nullity of the combinatorial geometry $G = G(E)$ whose points are the atoms of L.

If G has nullity 0, it is a Boolean algebra. Hence by Example, 7.1.5, $\mu_G(0,1) = (-1)^{|E|} = (-1)^{r(L)}$, whence (7.5) is immediate. This case includes lattices with rank 0 or 1.

If G has positive nullity, it is not a Boolean algebra. Hence there is a point e which is not an isthmus. By induction on rank, $(-1)^{r(G/e)}\mu(G/e) > 0$. By induction on nullity, $(-1)^{r(G-e)}\mu(G-e) > 0$. By Theorem 7.1.7,

$$(-1)^{r(G)}\mu(G) = (-1)^{r(G/e)}\mu(G/e) + (-1)^{r(G-e)}\mu(G-e),$$

which is positive by the previous observations. Thus we have the theorem. $\qquad\square$

7.1.9. Corollary. (*Brylawski 1972, Theorem 4.2 and Corollary 4.3*) *The magnitude of the Möbius invariant of a matroid satisfies*

$$|\mu(M)| = |\mu(M-e)| + |\mu(M/e)|$$

if $e \in E$ is neither an isthmus nor a loop, and

$$|\mu(M_1 \oplus M_2)| = |\mu(M_1)|\,|\mu(M_2)|. \qquad\square$$

The last result on μ is an expansion formula which will be needed to prove Stanley's modular-element factorization of the characteristic polynomial, Theorem 7.2.5 below.

7.1.10. Lemma. *Let x be a fixed element of the lattice L and let $v \in L$. Then*

$$\mu(0,v) = \sum_{\substack{y\ z \\ y \leqslant x, z \wedge x = 0 \\ y \vee z = v}} \mu(0,y)\mu(0,z).$$

Proof. Let $f(v)$ denote the right-hand side. Then

$$\sum_{u \leqslant v} f(u) = \sum_{y \leqslant x \wedge v} \sum_{\substack{z \leqslant v \\ z \wedge x = 0}} \mu(0,y)\mu(0,z)$$

$$= \delta(0, x \wedge v) \sum_{\substack{z \leqslant v \\ z \wedge x = 0}} \mu(0,z).$$

Either $\delta(0, x \wedge v) = 0$, or else $x \wedge v = 0$ so that the z-sum ranges over all $z \leqslant v$ and consequently equals $\delta(0,v)$. Thus $\sum_{u \leqslant v} f(u) = \delta(0,v)$. Inverting this sum yields $f(v) = \mu(0,v)$, as desired. $\qquad\square$

7.2. The Characteristic Polynomial

The characteristic polynomial is the matroid analog of the chromatic polynomial of a graph. While it does not count proper colorings—indeed there is no way known to color a general matroid corresponding to vertex coloring of a graph—the characteristic polynomial has most of the algebraic properties of the chromatic polynomial and can for many examples be interpreted in an interesting way related to coloring (in the 'critical problem').

The *characteristic polynomial*[†] *of a matroid M* is defined to be

$$p(M;\lambda) = \sum_{F \in L} \mu_M(\emptyset, F) \lambda^{r(M) - r(F)},$$

where L denotes the lattice of flats of M. Clearly, $p(M;\lambda)$ is monic of degree $r(M)$ except when $\emptyset$ is not closed, in which case $p(M;\lambda) \equiv 0$. The coefficient of $\lambda^{r(M)-k}$ is known as the *k-th Whitney number of the first kind of M*, written $w_k(M)$ (cf. Chapter 8); thus

$$p(M;\lambda) = \sum_{k=0}^{r(M)} w_k(M) \lambda^{r(M)-k},$$

$$w_k(M) = \sum_{\substack{F \in L \\ r(F) = k}} \mu_M(\emptyset, F).$$

We also see that $\mu(M) = p(M;0)$. Because of this, many properties of the Möbius invariant are specializations of those of the characteristic polynomial.

We also define the *characteristic polynomial of a geometric lattice L*; it is

$$p(L;\lambda) = \sum_{x \in L} \mu_L(0, x) \lambda^{r(L) - r(x)}.$$

This polynomial is always monic of degree $r(L)$; its coefficients are the Whitney numbers (of the first kind) of L. (Frequently in the literature $p(M; \lambda)$ is defined to be $p(L; \lambda)$ where L is the lattice of M. This is adequate for simple matroids but our definition is better in general.)

From our knowledge of the Möbius function we get at once two useful results. Setting $W = \emptyset$ and summing over all $F \in L$ in Proposition 7.1.4:

7.2.1. Proposition. *The characteristic polynomial of the matroid $M = M(E)$ has the Boolean expansion*

$$p(M;\lambda) = \sum_{X \subseteq E} (-1)^{|X|} \lambda^{r(M) - r(X)}. \qquad \square$$

7.2.2. Example. *Uniform matroids, Boolean algebras, and circuits.* For the

[†]Also called the *Birkhoff* or *Poincaré polynomial*.

uniform matroid U_{rm} with $0 < r \leqslant m$ we have

$$p(U_{rm}; \lambda) = \sum_{k=0}^{r-1} (-1)^k \binom{m}{k} [\lambda^{r-k} - 1]$$

$$= (-1)^{r-1}(\lambda - 1) \sum_{j=0}^{r-1} (-\lambda)^j \binom{m-1}{r-1-j}.$$

In particular $p(B_m; \lambda) = (\lambda - 1)^m$ and

$$P(C_m; \lambda) = (\lambda - 1)\frac{(\lambda - 1)^{m-1} - (-1)^{m-1}}{\lambda}.$$

From Rota's sign theorem, Theorem 7.1.8:

7.2.3. Proposition. *If L is a geometric lattice, then the coefficients of $(-1)^{r(L)}p(L; 1 - \lambda)$ are all positive. In other words,*

$$|w_k(L)| = (-1)^k w_k(L). \qquad \square$$

And from Proposition 7.2.1 we can deduce an analog of Theorem 7.1.7 contained essentially in Brylawski (1972, Theorem 4.2).

7.2.4. Theorem. *The characteristic polynomial of a matroid $M = M(E)$ satisfies:*

(i) *the deletion-contraction rule: if $e \in E$ is not an isthmus,*

$$p(M; \lambda) = p(M - e; \lambda) - p(M/e; \lambda);$$

(ii) *the direct sum rule: if $M = M_1 \oplus M_2$,*

$$p(M; \lambda) = p(M_1; \lambda)p(M_2; \lambda).$$

Proof. Exercise. See also Section 7.4. $\qquad \square$

Theorem 7.2.4(ii) shows that some characteristic polynomials factor in an interesting way. We can find a second kind of factorization by setting $\lambda = 1$. From the definition of μ_M, $p(M; 1) = 0$ for every matroid M whose point set is not empty. Hence $\lambda - 1$ divides $p(M; \lambda)$. Both factorizations are special cases of a theorem due to Stanley.

7.2.5. Theorem. [*Modular factorization (Stanley 1971, Theorem 2)*] *If x is a modular element of the geometric lattice L, then*

$$p(L; \lambda) = p([0, x]; \lambda) \sum_{\substack{z \in L \\ z \wedge x = 0}} \mu(0, z)\lambda^{r(L) - r(x) - r(z)}.$$

Proof. The right-hand side equals

$$\sum_{y \leqslant x} \sum_{z \wedge x = 0} \mu(0, y)\mu(0, z)\lambda^{r(L) - r(y) - r(z)}. \qquad (7.6)$$

We now need a lemma. Recall that $(v, w)M$ means v and w are a modular pair in L.

7.2.6. Lemma. *If* $(v, w)M$ *and* $v \wedge w \leqslant u \leqslant v$, *then* $(u, w)M$.

Proof of Lemma. We want to prove that

$$u \wedge (w \vee t) = (u \wedge w) \vee t \quad \text{for all} \quad t \leqslant u.$$

The inequality $\geqslant$ is a lattice identity, so it suffices to prove $\leqslant$. We have, by the assumption $(v, w)M$,

$$v \wedge (w \vee t) = (v \wedge w) \vee t \quad \text{for all} \quad t \leqslant v.$$

Note that $v \wedge w = u \wedge w$ and $v \geqslant u$. Hence

$$u \wedge (w \vee t) \leqslant (u \wedge w) \vee t \quad \text{for all} \quad t \leqslant v,$$

which is stronger than what we need. $\square$

In the theorem, $(x, z)M$ because x is a modular element; and $x \wedge z = 0 \leqslant y \leqslant x$. The lemma implies $(y, z)M$, whence

$$r(y) + r(z) = r(y \vee z) + r(y \wedge z) = r(y \vee z).$$

Thus (7.6) equals

$$\sum_{y \leqslant x} \sum_{z \wedge x = 0} \mu(0, y)\mu(0, z)^{r(L) - r(y \vee z)}$$

$$= \sum_{v \in L} \lambda^{r(L) - r(v)} \sum_{\substack{y \leqslant x \\ y \vee z = v}} \sum_{z \wedge x = 0} \mu(0, y)\mu(0, z).$$

The inner double sum equals $\mu(0, v)$ by Lemma 7.1.10. Thus we have the theorem. $\square$

To see that Stanley's theorem includes the direct-sum factorization, suppose $M = M_1 \oplus M_2$. Then $L = L(M) = L_1 \times L_2$. In L the element $x = (1_1, 0)$ is modular; moreover $z \wedge x = 0$ if and only if $z \in \{0\} \times L_2$. Thus in Stanley's theorem the first factor is $p(L_1; \lambda)$ and the second is $p(L_2; \lambda)$.

We can use Theorem 7.2.5 to determine the cofactor of $\lambda - 1$ in $p(L; \lambda)$. Let a be any atom of L; then $p([0, a]; \lambda) = \lambda - 1$. Let $L(a) = L - [a, 1]$; then $L(a)$ is an ideal in L. Define

$$p(L(a); \lambda) = \sum_{z \in L(a)} \mu_L(0, z)\lambda^{r(L) - 1 - r(z)}.$$

Since any atom is a modular flat, we have:

7.2.7. Corollary. *Let a be any atom in the geometric lattice L. Then $p(L; \lambda) = (\lambda - 1)p(L(a); \lambda)$.* $\square$

7.2.8. Corollary. *The polynomial $p(L(a); \lambda)$ is the same for every atom $a \in L$.*

□

7.2.9. Proposition. [*Brylawski 1971, Theorem 6.16(v)*] *Let M be the parallel connection of M_1 and M_2 with respect to the basepoint p, and assume p is not a loop in either M_1 or M_2. Then*

$$p(M; \lambda) = \frac{p(M_1; \lambda)p(M_2; \lambda)}{\lambda - 1}.$$

Proof. The assumptions imply that p is not a loop in M either. Hence by Corollary 7.2.7 the proposition is equivalent to the assertion that

$$p(L(p); \lambda) = p(L_1(p); \lambda)p(L_2(p); \lambda), \tag{7.7}$$

where L_i is the lattice of flats of M_i, provided that $\varnothing$ is closed in M_1 and M_2, which we may clearly assume. In full, (7.7) says

$$\sum_{\substack{F \in L \\ p \notin F}} \mu_L(0, F) \lambda^{r(L) - 1 - r(F)}$$

$$= \sum_{\substack{F_1 \in L_1 \\ p \notin F_1}} \sum_{\substack{F_2 \in L_2 \\ p \notin F_2}} \mu_{L_1}(0, F_1) \mu_{L_2}(0, F_2) \lambda^{r(L) - 1 - r(F_1) - r(F_2)},$$

since $r(L) - 1 = [r(L_1) - 1] + [r(L_2) - 1]$. By Brylawski (1971, Proposition 5.11) (see White 1986, Chapter 9), the flats F not containing p are precisely the unions $F_1 \cup F_2$ of flats of M_1 and M_2 where $p \notin F_1$ and $p \notin F_2$, and $[0, F] \cong [0, F_1] \times [0, F_2]$. So $r(F) = r(F_1) + r(F_2)$ and $\mu(0, F) = \mu(0, F_1)\mu(0, F_2)$, which is just what we need to prove the equation. □

7.3. The Beta Invariant

An informative number associated with a matroid is Crapo's beta invariant. With it one can decide whether a matroid is connected and whether it comes from a series-parallel network. The invariant can sometimes also establish that two matroids are not dual.

The *beta invariant* of the matroid $M = M(E)$, whose lattice of flats is L, is defined by

$$\beta(M) = (-1)^{r(M) - 1} \frac{d}{d\lambda} p(M; 1),$$

which equals $(-1)^{r(M) - 1} \sum_F \mu_M(\varnothing, F)[r(M) - r(F)]$, so that

$$\beta(M) = (-1)^{r(M)} \sum_{F \in L} \mu_M(\varnothing, F) r(F).$$

In view of Proposition 7.2.1 we could equally well define

$$\beta(M) = (-1)^{r(M)} \sum_{X \subseteq E} (-1)^{|X|} r(X),$$

as Crapo (1967) did when introducing the invariant. Some simple properties of β are summarized in Proposition 7.3.1.

7.3.1. Proposition. *Let $M = M(E)$, $L =$ the lattice of flats of M.*
 (a) *If M has no loops, $\beta(M)$ depends only on L.*
 (b) *$\beta(\text{isthmus}) = 1$.*
 (c) *$\beta(M) = 0$ if $E = \varnothing$ or if M contains a loop.*
 (d) *If $e \in E$ is not a loop,*

$$\beta(M) = (-1)^{r(M)-1} \sum_{\substack{F \in L \\ e \notin F}} \mu_M(\varnothing, F).$$

Proof. Exercise 7.17.   □

We define the beta invariant of a geometric lattice to be that of the underlying combinatorial geometry. Proposition 7.3.1 shows that $\beta(L) = 0$ if $r(L) = 0$, 1 if $r(L) = 1$.

The fundamental properties of β are those in Theorem 7.3.2.

7.3.2. Theorem. *(Crapo 1967) The beta invariant of the matroid $M = M(E)$ satisfies*
 (a) *$\beta(M) \geqslant 0$.*
 (b) *$\beta(M) > 0$ if and only if M is connected and is not a loop.*
 (c) *If $e \in E$ is neither an isthmus nor a loop,*

$$\beta(M) = \beta(M - e) + \beta(M/e).$$

 (d) *$\beta(M^*) = \beta(M)$ except when M is an isthmus or a loop.*

Proof of (c). Exercise 7.18.   □

Proof of (a). Exercise 7.18.   □

Proof of (b). By Exercise 7.18, $\beta(M) = 0$ if M is disconnected. We have to prove $\beta(M) > 0$ if M is connected and not a loop. If M is connected and $|E| \geqslant 3$, then [White 1986, Proposition 7.69 (1)] for every element e either M/e is connected or $M - e$ is connected. Then, by induction, since $|E - e| \geqslant 2$, either $\beta(M/e) > 0$ or $\beta(M - e) > 0$, hence by (c) and (a) we have $\beta(M) > 0$. The cases with $|E| \leqslant 2$ are easily checked to start the induction.   □

Proof of (d). Since M is disconnected if and only if M^* is, the disconnected case follows from (b). We may now assume that M is connected and $|E| \geqslant 2$.

Since no point is an isthmus or a loop, we have from (c):

$$\beta(M) = \beta(M - e) + \beta(M/e),$$
$$\beta(M^*) = \beta(M^* - e) + \beta(M^*/e)$$
$$= \beta((M/e)^*) + \beta((M - e)^*).$$

Then (d) follows by induction on $|E|$ provided $|E - e| \geqslant 2$. But if $|E - e| = 1$, M and M^* are both isomorphic to the 2-point circuit; (d) follows.   □

One of the uses of the beta invariant is to characterize series-parallel networks. First we establish the behaviour of β under series and parallel connections.

7.3.3. Proposition. [*Brylawski 1971, Theorem 6.16 (vi)*] *Let $M = M(E)$ be the series or parallel connection of two matroids M_1 and M_2, each having at least two points, with respect to the basepoint p. Then $\beta(M) = \beta(M_1)\beta(M_2)$.*

Proof. Suppose M is the parallel connection. The proposition follows from Proposition 7.3.1(d), Corollary 7.2.7, and Proposition 7.2.9.

But if M is the series connection, M^* is the parallel connection of M_1^* and M_2^*; the result follows from the former case and Theorem 7.3.2(d).   □

7.3.4. Proposition. [*Brylawski 1971, Theorem 7.6(2)*] *M is the matroid of a series-parallel network if and only if it is not an isthmus and $\beta(M) = 1$.*

Proof. The smallest series-parallel matroid is the 2-point circuit C_2. By Proposition 7.3.1, $\beta(C_2) = 1$. As White (1986, Chapter 6) shows, any series-parallel matroid is obtained from C_2 by a succession of parallel duplications of a point [which by Proposition 7.3.1(a) leave β unaltered] and dualizations [which do not change β due to Theorem 7.3.2(d)]. Hence $\beta(M) = 1$ if M is the matroid of a series-parallel network.

Conversely, suppose that $\beta(M) = 1$ and let $e \in E$, the point set of M. If $|E| = 1$, M must be an isthmus. Assuming now $|E| \geqslant 2$, M is connected [by Theorem 7.3.2 (b)] so Theorem 7.3.2(c) holds; since β is always a non-negative integer, we conclude that $\beta(M - e) = 0$ or $\beta(M/e) = 0$. Say the former: then $M - e = M(E_1) \oplus M(E_2)$; and $\beta(M) = \beta(M - E_2)\beta(M - E_1)$ by Proposition 7.3.3. So M is the series connection of two matroids with $\beta = 1$, which by induction on $|E|$ are series-parallel matroids. But then M is a series-parallel matroid.   □

Oxley [1982, Proposition (2.5)] extends Proposition 7.3.4 to larger values of β. He shows that, if $\beta(M) = k > 1$, then either M is a series-parallel extension of a 3-connected matroid with $\beta = k$ or M is a 2-sum of two matroids with $\beta < k$. See Oxley's paper for the definitions and proofs.

The beta invariant may be regarded as almost the Möbius inverse of the rank function. Let L be a geometric lattice; then the signed beta function $B(x) \equiv (-1)^{r(L)-r(x)}\beta(L/x)$ equals

$$\sum_{y \geqslant x} \mu_L(x, y) r(y).$$

Inverting,

$$r(x) = \sum_{y \geqslant x} B(y).$$

An expression essentially equivalent to this one appears in the cluster analysis of percolation processes on a graph (cf. Essam 1971, Sections 3.6–3.7).

7.4. Tutte–Grothendieck Invariance

The *rank generating function* of a matroid $M = M(E)$, introduced in Crapo (1970), is the two-variable polynomial

$$R(M; u, v) = \sum_{X \subseteq E} u^{r(M)-r(X)} v^{|X|-r(X)}.$$

The Boolean expansion theorems 7.1.5 and 7.2.1 amount to saying that $\mu(M)$ and $p(M; \lambda)$ are approximately specializations of $R(M; u, v)$; specifically,

$$\mu(M) = (-1)^{r(M)} R(M; 0, -1),$$
$$p(M; \lambda) = (-1)^{r(M)} R(M; -\lambda, -1).$$

Those observations and all the ideas of this section are based on Tutte (1947), where they were developed for graphs. Their extension to matroids is due to Crapo, Rota, and Brylawski.

The rank generating polynomial has an important property which generalizes Theorems 7.1.7 and 7.2.4. We need some definitions. An *invariant* of matroids is any function f of matroids which is the same for isomorphic matroids:

$$M \cong M' \quad \text{implies} \quad f(M) = f(M').$$

(We are only concerned, as usual, with finite matroids.) A *Tutte–Grothendieck invariant of matroids* is an invariant satisfying the direct-sum rule

$$f(M_1 \oplus M_2) = f(M_1) f(M_2)$$

and the deletion-contraction rule

$$f(M) = f(M - e) + f(M/e)$$

for each point e of M that is neither a loop nor an isthmus.

7.4.1. Proposition. *The rank generating function is a Tutte–Grothendieck invariant of matroids.*

Proof. Exercise 7.24. This result is implicit in Crapo (1970, Propositions 9 and 10), and is made explicit in Brylawski (1972).   □

Theorems 7.1.7 and 7.2.4 are special cases because any specialization of $R(M;u,v)$ is automatically a Tutte–Grothendieck invariant. The remarkable thing is that there is a converse.

7.4.2. Proposition. (*Brylawski 1972*) *If $f(M)$ is a Tutte–Grothendieck invariant of matroids, then it is an evaluation of $R(M;u,v)$. It is obtained by setting $u = f(\text{isthmus}) - 1$ and $v = f(\text{loop}) - 1$.*

Proof: Exercise 7.25.   □

This is a fundamental result but it still does not capture the essence of the characteristic polynomial. For that we need to define a *Tutte–Grothendieck invariant of geometries*. This is a matroidal Tutte–Grothendieck invariant with the additional property that

$$f(M) = f(G(M)) \quad \text{if} \quad M \text{ is loopless.}$$

7.4.3. Theorem. (*Brylawski 1972, Corollary 4.4*) *The invariant $(-1)^{r(M)}p(M;\lambda)$ is a Tutte–Grothendieck invariant of geometries. Moreover, it is a universal such invariant: if f is any such invariant, then $f(M) = (-1)^{r(M)}p(M;1 - f(\text{isthmus}))$.*

Proof. The geometric invariance of $(-1)^{r(M)}p(M;\lambda)$ follows from the definition of $p(M;\lambda)$ and from Theorem 7.2.4. Given f, in view of Proposition 7.4.2 it is enough to show that $f(B_1^*) = 0$. Let us consider $M = C_2$, the 2-point circuit. We have

$$f(B_1) = f(C_2) = f(C_2 - p) + f(C_2/p) = f(B_1) + f(B_1^*),$$

whence $f(B_1^*) = 0$.   □

7.5. Examples

Aside from the graphic matroids, chosen for their historical and motivating importance, our examples are of matroids whose characteristic polynomials are particularly simple in form because they belong to the class of 'supersolvable' geometries.

The chromatic polynomial. One of the *raisons d'être* of the characteristic polynomial, indeed its original motivation, is that it generalizes the chromatic polynomial of a graph. Let $\chi(\Gamma;\lambda)$ be the chromatic polynomial, $c(\Gamma)$ the number of components, and M the matroid of the graph Γ.

7.5.1 Proposition. $\chi(\Gamma;\lambda) = \lambda^{c(\Gamma)}p(M;\lambda)$.

This formula can be traced back to G.D. Birkhoff's paper of 1912, where it was stated (not for graphs, but for maps) in the form

$$\chi(\Gamma;\lambda) = \sum_{i=1}^{n} \lambda^i \sum_{k=0}^{n-i} (-1)^k c_k(0, n - i), \qquad (7.8)$$

n being the number of vertices and $c_k(0, n - i)$ the number of chains of length k from rank 0 to rank $n - i$ in the lattice of contractions of Γ [isomorphic to $L(M)$]. The equivalence of (7.8) with Proposition 7.5.1 is a consequence of Philip Hall's theorem (Exercise 7.13) and the fact that $c(\Gamma) = n - r(M)$ (White 1986, Chapter 6).

Proposition 7.5.1 is often proved by observing that $\chi(\Gamma;\lambda)/\lambda^{c(\Gamma)}$ is, like $p(M;\lambda)$, a Tutte–Grothendieck invariant of graphic matroids, comparing the two for a loop and an isthmus, and deducing their equality. But that approach does not explain the appearance of the Möbius function. For that it is better to carry out a proof by Möbius inversion (due essentially to Whitney 1932).

Proof. Let γ be any coloring of Γ in λ colors, whether proper or not, and let $I(\gamma)$ be the set of edges which are improperly colored, that is, $e \in I(\gamma)$ if and only if $I(\gamma)$ gives the same values to the two end points of e. It is easy to see that $I(\gamma)$ is closed in the graphic matroid M. Let L be the lattice of closed sets, and let $v(F) =$ the number of colorings γ for which $I(\gamma) = F$. Clearly

$$\sum_{F \in L} v(F) = \lambda^n.$$

More generally,

$$\sum_{F \geqslant F'} v(F) = \lambda^{c(F')},$$

since the colorings γ being counted, those which are improper on F' at least, have to be constant on each component of F'. Inverting,

$$\sum_{F \geqslant F'} \mu(F', F)\lambda^{c(F)} = v(F').$$

Setting $F' = 0$, on the left we have $\lambda^{c(\Gamma)}p(M;\lambda)$ and on the right $\chi(\Gamma;\lambda)$. (The trivial case where $\varnothing$ is not closed can be handled separately.) $\qquad \square$

Supersolvable geometric lattices (Stanley 1972). A geometric lattice is *supersolvable* when it contains a complete chain of modular elements. For such a lattice the modular-element factorization theorem makes computation of the characteristic polynomial easy.

7.5.2. Proposition. (*Stanley 1971, p. 217; 1972, Theorem 4.1*) *Suppose L is a geometric lattice of rank r with a complete chain $0 < x_1 < x_2 < \cdots < x_r = 1$ consisting of modular flats. Let $N_k =$ the number of atoms which are $\leqslant x_k$ but*

$\nleqslant x_{k-1}$. *Then*

$$p(L; \lambda) = (\lambda - 1)(\lambda - N_2)(\lambda - N_3) \cdots (\lambda - N_r),$$
$$\mu(L) = (-1)^r N_2 N_3 \cdots N_r,$$
$$\beta(L) = (N_2 - 1)(N_3 - 1) \cdots (N_r - 1),$$
$$w_k(L) = (-1)^k \sigma_k(1, N_2, N_3, \ldots, N_r),$$

where σ_k is the k-th elementary symmetric function.   □

One class of supersolvable geometric lattices is the Boolean algebras, or lattices of free matroids. Less trivial examples appear below.

Partitions. The partition lattice Π_n has characteristic polynomial

$$p(\Pi_n; \lambda) = (\lambda - 1)(\lambda - 2) \cdots (\lambda - n + 1). \qquad (7.9)$$

The coefficient of λ^k in $\lambda(\lambda - 1) \cdots (\lambda - n + 1)$ is by definition the Stirling number $s(n, k)$ of the first kind [hence the name 'Whitney number of the *first* kind' for w_{n-k}, since by (7.9), the Whitney number $w_{n-k}(\Pi_n)$ equals the Stirling number $s(n, k)$].

Projective geometries. Consider L_q^n, the lattice of subspaces of the n-dimensional vector space over $GF(q)$, equivalently of the projective geometry PG_q^{n-1}. Let σ_k denote the k-th elementary symmetric function.

7.5.3. Proposition. *We have*

$$p(L_q^n; \lambda) = (\lambda - 1)(\lambda - q)(\lambda - q^2) \cdots (\lambda - q^{n-1}),$$
$$\mu(L_q^n) = (-1)^n q^{\binom{n}{2}},$$
$$w_k(L_q^n) = (-1)^k \sigma_k(1, q, q^2, \ldots, q^{n-1})$$

$$= (-1)^k \sum_{0 \leqslant i_1 < i_2 < \cdots < i_k < n} q^{i_1 + i_2 + \cdots + i_k},$$

$$\beta(L_q^n) = (q - 1)(q^2 - 1) \cdots (q^{n-1} - 1).$$

Proof. Excercise.   □

From this proposition it is possible to compute $W_k(L_q^n)$, the number of distinct $(k - 1)$-dimensional subspaces of PG_q^{n-1}. See Exercise 7.31 (c).

7.6. The Critical Problem

The problem of coloring a graph is solved by finding the smallest positive integral argument such that $\chi_\Gamma(\lambda) > 0$. In the matroidal analog introduced by Crapo and Rota (1970), colors become vectors over the finite field of order q and one must find the smallest positive integral exponent d for which $p(M;q^d) > 0$.

The problem concerns a set E of vectors in the n-dimensional vector space K^n over $K = GF(q)$. Let $M(E)$ be the linear dependence matroid of E and $L(E)$ the lattice of flats of $M(E)$. A set of linear functionals $f_i: K^n \to K$ is said to *distinguish* E if for each point $p \in E$ some functional is non-zero on p; or in other words the intersection of the hyperplanes $\text{Ker} f_i$ is disjoint from E. The *critical problem* is to find the smallest size of a distinguishing set for E. We call this number c the *critical exponent* of E.

7.6.1. Theorem. [*Critical Theorem* (*Crapo and Rota 1970, Theorem 16.1*)] *Let* $E \subseteq K^n$, $m = \dim E$, *and* $d \geq 0$. *The number of (ordered) d-tuples of linear functionals which distinguish E (equivalently, the number of linear mappings $f: K^n \to K^d$ whose kernel avoids E) is equal to* $(q^d)^{n-m} p(M(E);q^d)$.

The most important conclusion to be drawn is that *the critical exponent of E is the smallest non-negative integer c such that $p(M(E);q^c) > 0$. We also see:*

7.6.2. Corollary. *Let E be a non-empty subset of a linear (or projective) space over $GF(q)$, not containing the zero vector. Then there is an integer $c > 0$ such that $p(M(E);q^d) = 0$ if $0 \leq d < c$ but $p(M(E);q^d) > 0$ for all $d \geq c$.* □

Proof of Theorem. The proof is similar to that of Proposition 7.5.1. First we observe that, given $X \subseteq K^n$ with $\dim X = e$, the number of linear mappings $f: K^n \to K^d$ whose kernel contains X is $q^{d(n-e)}$; for if we extend X to a spanning set by adjoining $p_{e+1}, \ldots, p_n$, we get such an f by setting $f|X = 0$ and choosing $f(p_i)$ arbitrarily from among the q^d vectors of K^d for $i = e+1, \ldots, n$. Now for each $F \subseteq K^n$, let us write $v(F)$ for the number of linear $f: K^n \to K^d$ such that $E \cap \text{Ker} f = F$. Obviously $E \cap \text{Ker} f$ is closed in $M(E)$, so we have for each $X \in L(E)$:

$$\sum_{F \geq X} v(F) = (q^d)^{n-e}.$$

After Möbius inversion and setting $X = 0 = \text{cl} \varnothing$,

$$\sum_{F \in L(E)} \mu_E(0,F)(q^d)^{n-\dim F} = v(0).$$

But the left-hand side equals $(q^d)^{n-m} p(L(E);q^d)$. Modulo obvious remarks about the case where $\varnothing$ is not closed, this is the theorem. □

The case of critical exponent 1 is easy to interpret geometrically. A combinatorial geometry is *affine* if it is isomorphic to the affine dependence matroid of a point set in an affine geometry AG_q^n. (We regard q as fixed.) A subset of PG_q^n is *affinely embedded* if it lies in the complement of a hyperplane.

Clearly $M(E)$ is affine if E is affinely embedded. We have the following converse and criterion. (The criterion, i.e., $c = 1$, is Theorem 16.2 of Crapo and Rota 1970. It is a q-analog of the Two-Color Theorem of graph theory; see below.)

7.6.3. Corollary. *Let* $E \subseteq PG_q^n$. *The following are equivalent:*
  (i) *E is affinely embedded.*
  (ii) *$M(E)$ is affine.*
  (iii) *E has critical exponent 1.*

Proof. Exercise.   □

The Critical Theorem shows in principle how to find the critical number (although drawing conclusions in specific cases is another matter!), but what it counts is not very geometrical. One can deduce more complicated expressions for the number of d-tuples of hyperplanes (as distinct from functionals) which distinguish E and the number of e-dimensional subspaces which avoid E.

7.6.4. Corollary. *Let* $E \subseteq K^n$ *and let* $m = \dim E$. *(Or let* $E \subseteq PG_q^{n-1}$ *and* $m = 1 + \dim E$.) *The number of d-tuples of hyperplanes which distinguish E (i.e., whose intersection avoids E) is equal to*

$$(q-1)^{-d} \sum_{e=0}^{d} (-1)^{d-e} (q^e)^{n-m} p(M(E); q^e)$$

$$= \sum_{k=0}^{m} w_k(M(E)) \left(\frac{q^{n-k} - 1}{q-1} \right)^d.$$

Proof. Let κ_d (respectively v_d) be the number of d-tuples of hyperplanes (respectively functionals) that distinguish E. We have to take account of two factors: some functionals are 0 (not corresponding to any hyperplane), and one hyperplane corresponds to $q - 1$ functionals.

We can obtain all d-tuples f of functionals that distinguish E in the following way. First we choose $e = 0, 1, \ldots,$ or d (e will be the number of non-zero functionals in f) and one of the κ_e e-tuples of distinguishing hyperplanes, $h = (h_1, h_2, \ldots, h_e)$. Next for each h_j we pick one of the $q - 1$ functionals g_j with h_j for kernel. Then we pick e indices, $1 \leq i_1 < i_2 < \cdots < i_e \leq d$, and we let $f_{i_j} = g_j$, but $f_i = 0$ if i is not one of the selected indices. This determines f, and since the e-tuple h is recoverable from f, we obtain in this way all possible f. So

$$v_d = \sum_{e=0}^{d} \kappa_e (q-1)^e \binom{d}{e}.$$

Inverting this binomial relation,

$$\kappa_d = (q-1)^{-d} \sum_{e=0}^{d} (-1)^e \binom{d}{e} v_e.$$

We obtain the value of v_e from the Critical Theorem.

The alternate form of κ_d arises upon expanding $p(M(E);q^d) = \sum w_k(M(E))(q^{m-k})^d$ and rearranging the sum. $\square$

Finally, we have the most geometrical version of the Critical Theorem. But first this lemma.

7.6.5. Lemma. *Let x be an $(n-e)$-dimensional subspace of K^n. The number of linear mappings $f: K^n \to K^d$ whose kernel is x equals*

$$(q^d - 1)(q^d - q)\cdots(q^d - q^{e-1}),$$

interpreted as 1 if $e = 0$.

Proof. There is a one-to-one correspondence between such f and the mappings $\bar{f}: K^n/x \to K^d$ with zero kernel. To count the latter is a critical problem: we want the number of mappings $\bar{f}: K^e \to K^d$ whose kernel avoids $E = K^e - \{0\}$. This number is $p(M(E);q^d)$. But we know that polynomial from Proposition 7.5.3 since $L(E) \cong L_q^e$. So we have the lemma. $\square$

7.6.6. Proposition. (*Dowling 1971, Theorem 2, p. 220*) *Let $E \subseteq K^n$ have dimension m. The number of $(n-d)$-dimensional subspaces of K^n not meeting E is equal to*

$$\sum_{e=0}^{d} \frac{(-1)^e(q^{d-e})^{n-m}p(M(E);q^{d-e})}{(q^e-1)(q^{e-1}-1)\cdots(q-1)(q^{d-e}-1)(q^{d-e}-q)\cdots(q^{d-e}-q^{d-e-1})}.$$

Notice that the terms with $e > d - c, c$ the critical exponent of E, are all 0.

Proof. Let σ_{n-d} denote the number of $(n-d)$-dimensional subspaces avoiding E. We will set up and solve a recurrence for σ_{n-d}.

Each subspace counted by σ_{n-e} is the kernel of the number of mappings $K^n \to K^d$ given by Lemma 7.6.5. So the total number of mappings $K^n \to K^d$ whose kernels avoid E is given by

$$\sum_{e=0}^{d} \sigma_{n-e} \prod_{i=0}^{e-1} (q^d - q^i).$$

The number of such mappings is also given by the Critical Theorem; thus we have

$$(q^d)^{n-m}p(M(E);q^d) = \sum_{e=0}^{d} \sigma_{n-e} \prod_{i=0}^{e-1} (q^d - q^i).$$

The trick is to rewrite this as an identity involving the *Gaussian coefficients*,

$$\begin{bmatrix} d \\ e \end{bmatrix} = \frac{(q^d - 1)(q^{d-1} - 1)\cdots(q^{d-e-1} - 1)}{(q^e - 1)(q^{e-1} - 1)\cdots(q - 1)},$$

which is to be proved in Exercises 7.5 and 7.31 to equal the number of

e-dimensional subspaces of K^d. That is, we want to prove

$$(q^d)^{n-m}p(M(E); q^d) = \sum_{e=0}^{d} \sigma_{n-e}\begin{bmatrix} d \\ e \end{bmatrix} \prod_{i=0}^{e-1} (q^e - q^i). \qquad (7.10)$$

Equation (7.10) has the form

$$a_d = \sum_{e=0}^{d} b_e \begin{bmatrix} d \\ e \end{bmatrix}, \qquad (7.11)$$

valid for all $d \geqslant 0$. We wish to solve for b_e. That we can do by defining, for $x \in L_q^n$ with dim $x = d$,

$$a(x) = a_d \quad \text{and} \quad b(x) = b_d.$$

Now (7.11) can be written

$$a(x) = \sum_{y \leqslant x} b(y),$$

which by Möbius inversion in L_q^n becomes

$$b(x) = \sum_{y \leqslant x} a(y)\mu(x, y).$$

The interval $[x, y]$ being a projective geometry, its Möbius invariant is given by Proposition 7.5.3; converting back to the notation of (7.11) we have

$$b_d = \sum_{e=0}^{d} a_{d-e}(-1)^e q^{\binom{e}{2}} \begin{bmatrix} d \\ e \end{bmatrix}.$$

The result of inverting (7.10) in this fashion is

$$\sigma_{n-d} \prod_{i=0}^{d-1} (q^d - q^i) = \sum_{e=0}^{d} (-1)^e q^{\binom{e}{2}} \begin{bmatrix} d \\ e \end{bmatrix} (q^{d-e})^{n-m} p(M(E); q^{d-e}).$$

Isolating σ_{n-d} and simplifying yields the result. □

7.6.7. Corollary. *The largest dimension of a substance of K^n not meeting E is $n - c$, where c is the critical exponent of E.* □

7.6.8. Example. *Independent sets.* Any independent set of points has critical exponent 1 and therefore lies in the complement of a hyperplane in K^n.

Graph coloring as a critical problem. Since a graphic matroid can be represented by vectors over any field, it has a critical problem for each prime power q. Let Γ, a graph with n vertices, be represented by the vector set $E(\Gamma) \subseteq K^n$, where $K = GF(q)$, in the usual way: vertex v_i corresponds to the i-th coordinate and an edge e_{ij} corresponds to the vector $p_i - p_j$ (or $p_j - p_i$), $\{p_i\}$ being the standard basis of K^n. Each linear mapping $f: K^n \to K^d$ corresponds to a coloring of Γ by K^d, that is, a map $\gamma: V(\Gamma) \to K^d$ defined by $\gamma(v_i) = f(p_i)$; and conversely each γ determines one linear mapping f. Moreover, f distinguishes $E(\Gamma)$ if and only if, for each edge e_{ij} of Γ, $f(p_i - p_j) \neq 0$; in other words, γ is a proper coloring. So in the graphic case the Critical Theorem says that $\chi(\Gamma; q^d)$ *is the number of proper colorings of Γ by vectors in K^d*; the critical exponent is the smallest dimension d for which there is a proper coloring by K^d. (One should now reread Corollary 7.6.3 as a q-color theorem!)

The most interesting case is the binary one, for the statement: *the critical exponent of a planar graph, over $GF(2)$, is at most 2*, is the Four-Color Theorem. An aim of Crapo and Rota in formulating the critical problem was to put the Four-Color Problem in a general setting which might lead to techniques powerful enough to solve it and other problems of the type. ('The fact that the problem of coloring a graph was the first historically to arise, was a distressing accident, which prevented it from being studied at that level of generality which has been found indispensible in solving most problems of mathematics.') It must be admitted that this hope has not yet been realized, although it is undoubtedly worthy of continued pursuit.

Linear codes and the critical problem. Another example was pointed out by Dowling (1971). A *linear code* in K^n with *distance* d is a linear subspace whose non-zero vectors have minimum weight d. (The *weight* of a vector is the number of non-zero coordinates.) The problem of linear coding theory is to find large codes with given dimension and given (or bigger) distance. Suppose we let

$$E_\delta = \{p \in K^n: \ 0 < \text{wt}(p) \leqslant \delta\},$$

and c_δ = the critical exponent of E_δ. Then a code with distance $> \delta$ is merely a subspace avoiding E_δ; by the Critical Theorem the largest dimension of such a subspace is $n - c_\delta$ and its size is q^{n-c_δ}. So if we can calculate $p(M(E_\delta); \lambda)$ we will know the maximum size of a linear code with distance $> \delta$.

This is a difficult calculation in general, although easy when $\delta = 1$ (Exercise). Dowling accomplished the calculation for $\delta = 2$. Then $L(E_2)$ is the *Dowling lattice* $Q_n(K^*)$ of the multiplicative group K^* of K (Dowling 1973a; for the Dowling lattices of any finite group see Dowling 1973b). The characteristic polynomial of $Q_n(K^*)$ evaluated at q^d equals

$$(q-1)^n \left(\frac{q^d - 1}{q - 1} \right) \left(\frac{q^d - 1}{q - 1} - 1 \right) \cdots \left(\frac{q^d - 1}{q - 1} - n + 1 \right),$$

by Dowling's results. Thus c_2 is the integer such that

$$2^{c_2 - 1} \leqslant n < 2^{c_2}, \quad \text{if} \quad q = 2, \text{ or}$$

$$\frac{q^{c_2 - 1} - 1}{q - 1} < n \leqslant \frac{q^{c_2} - 1}{q - 1}, \quad \text{if} \quad q > 2.$$

Then we know the maximum size of a linear code over $GF(q)$ that corrects one error (which is what $\delta = 2$ signifies). This problem was what led Dowling to investigate his lattices and thence to the theory of Dowling (1973b).

Unfortunately, for $\delta \geqslant 3$ this approach does not succeed. The reason is roughly that $M(E_2)$ is essentially graphic, as one can see from the presentation given in Dowling (1973a, p. 109); moreover, it is supersolvable. For larger δ, $M(E_\delta)$ is no longer graphic; the techniques to calculate its characteristic polynomial have not been discovered. This is one of the important open problems in matroid theory.

A different connection between linear codes and the critical problem and also one between codes and the rank generating function (Section 7.4) are developed in Greene (1976).

Exercises

7.1.   Prove Proposition 7.1.1 using the recursive definition, equations (7.1)–(7.3).

7.2.   Prove Proposition 7.1.2: first from the definition of μ, then using the incidence algebra.

7.3.   (a) Evaluate $\mu(U_{rm})$ (Example 7.1.5).

   (b) Find and factor the characteristic polynomial of the m-point line U_{2m}.

   (c) Deduce Example 7.2.2 from Proposition 7.2.1. Calculate the Whitney numbers of the first kind of U_{rm}, B_m, C_m.

7.4.   (a) For the partition lattice Π_n, evaluate $\mu(\Pi_n)$:

   (i) from the definition (7.1.) for $n = 4$;

   (ii) from the alternative recurrence (Proposition 7.1.2) (Frucht and Rota 1963).

   (iii) Deduce that, if $\pi \leqslant \tau$ in Π_n and π partitions n_i different blocks of τ into i parts each for $i = 1, 2, 3, \ldots$, then

$$\mu(\pi, \tau) = (-1)^{|\pi| - |\tau|} (1!)^{n_2} (2!)^{n_3} (3!)^{n_4} \ldots .$$

   (Schützenberger 1954)

   (b) Deduce a formula for $p(\Pi_n; \lambda)$ from the definition of the characteristic polynomial and Exercise 7.4(a). What are the Whitney numbers $w_k(\Pi_n)$?

7.5.   (a) For the lattice L_q^n of the projective geometry PG_q^{n-1} of dimension $n - 1$ over $GF(q)$, evaluate $\mu(L_q^n)$. Then calculate $\mu(L(AG_q^{n-1}))$, where AG_q^{n-1} is the affine geometry. (You may express the result in terms of the numbers $W_k(L)$ of rank k subspaces of PG_q^{n-1}.)

   (b) Find the characteristic polynomial and Whitney numbers of the first kind of L_q^n, based on your solution to (a). Do the same for $L(AG_q^{n-1})$.

7.6.   Let V_n consist of all the points in the real affine space $AG^n(\mathbb{R})$ with coordinates ± 1; we call this the *verticial hypercube*. If $n = 3$, it is called the real affine cube. For the geometric lattice of its affine dependence matroid, compute the Möbius invariant and the characteristic polynomial when $n \leqslant 3$. The general problem is unsolved, difficult, and important. It would yield an exact formula for the number of threshold switching functions of n variables (Winder 1966; Zaslavsky 1975, Section 5F).

7.7. (a) Let $\hat{D}_n$ consist of all the points in the real vector space $\mathbb{R}^n$ with exactly two non-zero coordinates, whose values are in the set $\{+1, -1\}$. Let $\hat{B}_n$ be $\hat{D}_n$ with the unit basis vectors adjoined. Let L denote the lattice of the linear dependence matroid. Compute $\mu(L(\hat{D}_n))$ and $\mu(L(\hat{B}_n))$ for $n \leqslant 3$, then $n = 4$ if time allows. [For general n, see Zaslavsky (1981).]

(b) Like Exercise 7.5(b) but for $\hat{D}_n$ and $\hat{B}_n$. *Hint*: $p(L(\hat{B}_n); \lambda) = (\lambda - 1)(\lambda - 3) \cdots (\lambda - 2n + 1)$.

7.8. Prove Proposition 7.1.3.

7.9. Prove Proposition 7.1.4. *Hints*: For the case $W \notin L$, factor the sum. For $W \in L$, define the function

$$\phi(W, F) = \sum_{\substack{W \subseteq X \subseteq F \\ \text{cl} X = F}} (-1)^{|X - W|}$$

and employ the incidence algebra.

7.10. Prove Theorem 7.1.7 (ii). *Hint*: Use Proposition 7.1.4.

7.11. Prove that the sum $i_0 - i_1 + i_2 - \cdots \pm i_r$, where i_k is the number of independent sets of rank k in a matroid M of rank r, equals zero if and only if M has an isthmus. *Hint*: Use Proposition 7.1.4.

7.12. Prove Rota's sign theorem, Theorem 7.1.8, from Weisner's theorem, Proposition 7.1.6. (Rota 1964)

7.13. [Philip Hall's Theorem: Hall 1936, (2.21); Rota 1964, Proposition 6, p. 346.] For $x, y \in P$ and $i \geqslant 0$, let $c_i(x, y)$ be the number of chains $x = x_0 < x_1 < \cdots < x_i = y$ of length i from x to y. Let
$$\phi(x, y) = c_0(x, y) - c_1(x, y) + c_2(x, y) - c_3(x, y) + \cdots.$$
Prove that $\mu(x, y) = \phi(x, y)$.

7.14. Prove Theorem 7.2.4 in a manner analogous to the proof of Theorem 7.1.7.

7.15. If $x \in L$ is modular, $L(x) = \{y \in L: y \wedge x = 0\}$, and $p(L(x); \lambda) = \sum \{\mu(0, y) \lambda^{r(L) - r(x) - r(y)}: y \in L(x)\}$, is $(\lambda - 1)p(L(x); \lambda)$ always the characteristic polynomial of a matroid? (Brylawski 1975, Section 7)

7.16. Discover and prove an analog of Proposition 7.2.9 for the generalized parallel connection of M_1 and M_2 along a common modular flat F (Brylawski 1975, Section 5; see White 1986, Chapter 9. *Hint*: Remember Stanley's theorem, Theorem 7.2.5 (Brylawski 1975, Theorem 7.8)).

7.17. Prove Proposition 7.3.1.

7.18. Prove Theorem 7.3.2(c), (a). Also show that $\beta(M) = 0$ when M is disconnected.

7.19. Use Theorem 7.3.2 to evaluate the beta invariant of (a) the m-point line U_{2m} and (b) Π_4.

7.20. Determine the value of $\beta(U_{rm})$. For which values of m and r is U_{rm} a series-parallel matroid?

7.21. Calculate β for the examples of Exercises 7.5, 7.6, 7.7. Is any one a series-parallel matroid?

7.22. Prove that $\beta(L) = (-1)^{r(L) - 1} \prod \{\mu(0, x): x \in L, x \not\geqslant a\}$ for every atom a of the geometric lattice L (Zaslavsky 1975, Section 7).

7.23. Calculate the rank generating function of U_{rm} directly from the definition.

7.24. Prove Proposition 7.4.1.

7.25. Prove Proposition 7.4.2. *Hint*: Use induction on the size of M.

7.26. Calculate $R(U_{rm}; u, v)$ from the Tutte–Grothendieck recurrence and the values for $m \leqslant 1$.

7.27. Compute $R(M(K_n); u, v)$, where $M(K_n)$ is the graphic geometry of the complete graph. How does your result, evaluated at $v = -1$ and $u = -\lambda$, compare with $p(\Pi_n; \lambda)$ from Exercise 7.4?

7.28.   (a) Prove that Π_n is supersolvable. (*Hint*: What about a partition with only one non-singleton block?) Deduce (7.9) and $\mu(\Pi_n)$ and $\beta(\Pi_n)$.

  (b) Prove (7.9) by graph theory *via* Proposition 7.5.1, since $\Pi_n \cong L(M(K_n))$.

  (c) Compare (7.9) to your answer to Exercise 7.4(b). What Stirling number identity is thereby proved?

7.29. Prove Proposition 7.5.1 by the Tutte–Grothendieck method.

7.30. Prove Proposition 7.5.2.

7.31.   (a) Prove   that $W_k(L_q^n) = \dfrac{(q^n - 1)(q^{n-1} - 1) \cdots (q^{n-k+1} - 1)}{(q^k - 1)(q^{k-1} - 1) \cdots (q - 1)}$ by   counting ordered bases.

  (b) Deduce p and μ of Proposition 7.5.3 from supersolvability (Stanley 1972, Example 4.2).

  (c) Compare with your results from Exercise 7.5(a). Deduce that
$$W_k(L_q^n) = \sum_{0 \leqslant j_1 \leqslant j_2 \leqslant \cdots \leqslant j_k \leqslant n-k} q^{j_1 + j_2 + \cdots + j_k}.$$

7.32.   (a) Calculate the critical exponent over $K = GF(q)$ of an m-point line $U_{2m}, 2 \leqslant m \leqslant q + 1$. Is U_{2m} affine in PG_q^{n-1}?

  (b) The same, for a circuit C_{r+1} of rank $r \geqslant 3$. Is C_{r+1} affine in PG_q^{n-1}? *Hint*: Almost always.

  (c) The same, for U_{rm} where $2 < r < m - 1$. (Assume U_{rm} is such that it embeds in PG_q^{n-1}.)

7.33. How many hyperplanes avoid a fixed non-empty set $E \subseteq K^n$? How many $(n-2)$-dimensional subspaces?

7.34. Prove Example 7.6.8. How many $(n-d)$-dimensional subspaces avoid a fixed basis?

7.35. If $q = p^e$, PG_p^n is a spanning subset of PG_q^n. What is its critical exponent?

7.36. Prove Corollary 7.6.3.

7.37. Deduce Corollary 7.6.7 directly from the Critical Theorem.

7.38. Calculate the critical exponent of the set $\hat{A}_n$ of all vectors in K^{n+1} with exactly two non-zero coordinates, one equal to $+1$ and the other equal to -1 (note that $+1 = -1$, if q is even). *Hint*: $M(\hat{A}_n) \cong M(K_{n+1})$, the complete-graph matroid.

7.39. What is the critical exponent of $\hat{B}_n$? (See Exercise 7.7. Assume q is odd. *Hint*: The matroid of $\hat{B}_n$ is the same for $K = \mathbb{R}$ and $K = GF(q)$ as long as q is odd.)

7.40.   (a) Calculate the critical exponent c_1 of E_1. What is the maximum size of a linear code with distance $\geqslant 2$ (a code that detects one error)?

  (b) Express compactly the maximum size of a linear code with distance $\geqslant 3$ (a code that corrects one error).

References

Aigner, M. (1979). Combinatorial Theory. *Grundlehren der Math. Wiss.* **234**, Springer-Verlag, Berlin–New York.

Birkhoff, G.D. (1912). A determinant formula for the number of ways of coloring a map. *Annals of Math.* (2) **14**, 42–6.

Brylawski, T.H. (1971). A combinatorial model for series-parallel networks. *Trans. Amer Math. Soc.* **154**, 1–22.

Brylawski, T.H. (1972). A decomposition for combinatorial geometries. *Trans Amer. Math. Soc.* **171**, 235–82.

Brylawski, T.H. (1975). Modular constructions for combinatorial geometries. *Trans Amer. Math. Soc.* **203**, 1–44.

Crapo, H.H. (1967). A higher invariant for matroids. *J. Comb. Theory* **2**, 406–17.

Crapo, H.H. (1970). The Tutte polynomial. *Aequationes Math.* **3**, 211–29.

Crapo, H.H. and Rota, G.C. (1970). *On the Foundations of Combinatorial Theory: Combinatorial Geometries* (preliminary edition). MIT Press, Cambridge, Mass.

Dowling, T.A. (1971). Codes, packings, and the critical problem, in *Atti del Convegno di Geometria Combinatoria e Sue Applicazioni (Perugia, 1970)*, pp. 209–24. Ist. Mat., Univ. di Perugia, Perugia, Italy.

Dowling, T.A. (1973a). A q-analog of the partition lattice, in *A Survey of Combinatorial Theory*, (J.N. Srivastava *et al.*, eds.), pp. 101–15. North-Holland, Amsterdam.

Dowling, T.A. (1973b). A class of geometric lattices based on finite groups. *J. Comb. Theory Ser. B* **14**, 61–86. Erratum, *ibid.* **15**, 211.

Essam, J.W. (1971). Graph theory and statistical physics. *Discrete Math.* **1**, 83–112.

Frucht, W.L. and Rota, G.C. (1963). La función de Möbius para paritiones de un conjunto. *Scientia (Valparaiso, Chile)* **122**, 111–15.

Greene, C. (1976). Weight enumeration and the geometry of linear codes. *Stud. Appl. Math.* **55**, 119–28.

Hall, P. (1936). The Eulerian functions of a group. *Quarterly J. Math.* **7**, 134–51.

Oxley, J.G. (1982). On Crapo's beta invariant for matroids. *Stud. Appl. Math.* **66**, 267–77.

Rota, G.C. (1964). On the foundations of combinatorial theory, I. Theory of Möbius functions. *Z. Wahrscheinlichkeitstheorie verw. Gebiete* **2**, 340–68.

Schützenberger, M.P. (1954). Contribution aux applications statistiques de la théorie de l'information, in *Publ. Inst. Statist. Univ. Paris 3*, Nos. 1–2, pp. 3–117.

Stanley, R.P. (1971). Modular elements of geometric lattices. *Algebra Universalis* **1**, 214–17.

Stanley, R.P. (1972). Supersolvable lattices. *Algebra Universalis* **2**, 197–217.

Tutte, W.T. (1947). A ring in graph theory. *Proc. Camb. Phil. Soc.* **43**, 26–40.

Weisner, L. (1935). Abstract theory of inversion of finite series. *Trans. Amer. Math. Soc.* **38**, 474–92.

White, N.L., ed. (1986). *Theory of Matroids.* Cambridge University Press.

Whitney, H. (1932). A logical expansion in mathematics. *Bull. Amer. Math. Soc.* **38**, 572–9.

Winder, R.O. (1966). Partitions of N-space by hyperplanes. *SIAM J. Appl. Math.* **14**, 811–18.

Zaslavsky, T. (1975). Facing up to arrangements: Face-count formulas for partitions of space by hyperplanes. *Mem. Amer. Math. Soc.* **1**, Issue 1; No. 154.

Zaslavsky, T. (1981). The geometry of root systems and signed graphs. *Amer. Math. Monthly* **88**, No. 2 (February).

8

Whitney Numbers

MARTIN AIGNER

8.1. Introduction

To every matroid we associate its lattice of flats and the lattices that arise in this way are known as geometric lattices and characterized (in the finite case) as (upper-) semimodular point lattices. Among the many numerical invariants of a finite geometric lattice such as rank, number of points, Möbius number, etc., we study in this chapter two sequences of numbers, the Whitney numbers of the first and of the second kind. After introducing these numbers and stating some of the basic results and problems in Section 8.2, we collect in Section 8.3 some material on the Möbius function that is needed in the subsequent Sections 8.4 and 8.5 where a survey of the known results on Whitney numbers is presented.

8.2. The Characteristic and Rank Polynomials

Let L be a finite graded lattice with 1, rank function r and $r(1) = r$. To L we associate two polynomials (over $\mathbb{Q}$), the *characteristic polynomial* $p(L; \lambda)$ and the *rank polynomial* $\rho(L; \lambda)$, defined by:

8.2.1. Definition.

(i) $p(L; \lambda) = \sum_{a \in L} \mu(0, a) \lambda^{r(1) - r(a)} = \sum_{k=0}^{r} w_k \lambda^{r-k},$

(ii) $\rho(L; \lambda) = \sum_{a \in L} \lambda^{r(1) - r(a)} = \sum_{k=0}^{r} W_k \lambda^{r-k},$

where μ denotes the Möbius function of L (cf. Chapter 7). The coefficients w_k and W_k, $0 \leqslant k \leqslant r$, are called the *Whitney numbers of the first and second kind*, respectively, i.e.,

$$w_k = \sum_{a: r(a) = k} \mu(0, a),$$

$$W_k = |\{a \in L : r(a) = k\}|.$$

It was shown in Chapter 7 that for geometric lattices the numbers $1 = w_0$, $w_1, w_2, \ldots, w_r$ are all non-zero with alternating sign whence

$$w_k^+ = (-1)^k w_k = (-1)^k \sum_{a:r(a)=k} \mu(0, a)$$

is positive for $k = 0, \ldots, r$. For ease of reference we will prove this result again in Corollary 8.3.5 below. For our purposes it is more convenient to consider the numbers w_k^+ which are sometimes called the *unsigned Whitney numbers of the first kind*.

It is the object of this chapter to study the sequences $(w_k^+ : 0 \leqslant k \leqslant r)$ and $(W_k : 0 \leqslant k \leqslant r)$ for geometric lattices. All lattices involved will be assumed to be finite.

Let us look at some basic examples first.

8.2.2. Examples.

(i) Let $L = B_n$ be the Boolean lattice of rank n. Then $\mu(0, a) = (-1)^{r(a)}$, and thus $w_k^+ = W_k = \binom{n}{k}$ for all k, $p(B_n; \lambda) = (\lambda - 1)^n$, $\rho(B_n; \lambda) = (\lambda + 1)^n$ (see Proposition 7.5.2).

(ii) Let $L = L(n, q)$ be the lattice of a projective space $PG(n-1, q)$ of dimension $n-1$ over $GF(q)$. Then $\mu(0, a) = (-1)^{r(a)} q^{\binom{r(a)}{2}}$ (see Proposition 7.5.3) and thus $w_k^+ = \begin{bmatrix} n \\ k \end{bmatrix} q^{\binom{k}{2}}$, $W_k = \begin{bmatrix} n \\ k \end{bmatrix}$ where $\begin{bmatrix} n \\ k \end{bmatrix}$ are the Gaussian coefficients, also denoted $\binom{n}{k}_q$ (cf. Aigner 1979, p. 78 and Exercise 7.31), yielding $p(L(n, q); \lambda) = \prod_{k=0}^{n-1}(\lambda - q^k)$, $\rho(L(n, q); \lambda) = \sum_{k=0}^{n} \begin{bmatrix} n \\ k \end{bmatrix} \lambda^{n-k}$.

(iii) Let $L = \prod_n$ be the partition lattice on n elements. A partition a has rank i iff it consists of $n - i$ blocks, hence $W_k = S_{n,n-k}$ where $S_{n,j}$ are the Stirling numbers of the second kind, thus $\rho(\prod_n; \lambda) = \sum_{k=1}^{n} S_{n,k} \lambda^{k-1}$. Furthermore we saw in (7.9) that $p(\prod_n; \lambda) = (\lambda - 1)(\lambda - 2) \cdots (\lambda - n + 1)$. With the usual expansion $\lambda(\lambda - 1) \cdots (\lambda - n + 1) = \sum_{j=1}^{n} s_{n,j} \lambda^j = \sum_{k=0}^{n-1} s_{n,n-k} \lambda^{n-k}$, we see that the numbers $w_k^+ = (-1)^k s_{n,n-k} = |s_{n,n-k}|$ are the absolute values of the Stirling numbers of the first kind. This is, in fact, the origin of the name 'Whitney numbers of the first and second kind'.

Let us gather some facts for an arbitrary graded lattice L of rank r.

$$\begin{aligned} &w_0 = W_0 = 1, \ w_1^+ = W_1 = \text{number of points of } L \\ &(= \text{elements covering } 0), \ w_r^+ = (-1)^r \mu(0, 1), \ W_r = 1. \end{aligned} \quad (8.1)$$

The set-up of the polynomials p and ρ as a convolution product suggests the introduction of the incidence algebra $A(L)$ over the ring $\mathbb{Q}[\lambda]$. (See Doubilet,

Rota, and Stanley 1972.) Let us define $\bar{r} \in A(L)$ by $\bar{r}(a,b) = \lambda^{r(b)-r(a)}$. Then, by Definition 8.2.1,

$$p(L;\lambda) = (\mu * \bar{r})(0,1),$$
$$\rho(L;\lambda) = (\zeta * \bar{r})(0,1) \tag{8.2}$$

where ζ is the ζ-function in $A(L)$ and $*$ denotes the convolution product. Since $\zeta^2(a,b) = |[a,b]|$ we conclude from (8.2) the following formula relating the polynomials p and ρ:

$$\rho(L;\lambda) = \sum_{a \in L} |[0,a]| p([a,1];\lambda). \tag{8.3}$$

Since ζ, μ and $\bar{r}$ are multiplicative functions we obtain a first decomposition theorem.

8.2.3. Proposition. *Let $L = L_1 \times L_2$ be a graded lattice. Then*

$$p(L;\lambda) = p(L_1;\lambda) \cdot p(L_2;\lambda),$$
$$\rho(L;\lambda) = \rho(L_1;\lambda) \cdot \rho(L_2;\lambda). \qquad \square$$

So far all our results are valid for arbitrary graded lattices. Now let us specialize to geometric lattices. An investigation of the Whitney numbers of lattices of small rank and of the main examples has led to the following conjectures.

Call a sequence $(v_0, v_1, \ldots, v_r)$ of non-negative real numbers *unimodal* if $v_i \geqslant \min(v_h, v_j)$ for all $0 \leqslant h \leqslant i \leqslant j \leqslant r$. In other words,

$$v_0 \leqslant v_1 \leqslant \cdots v_{i-1} \leqslant v_i = \cdots = v_j \geqslant v_{j+1} \geqslant \cdots \geqslant v_r. \tag{8.4}$$

The sequence is *logarithmically concave* if

$$v_k^2 \geqslant v_{k-1} v_{k+1} \quad \text{for} \quad 1 \leqslant k \leqslant r-1. \tag{8.5}$$

It is easily seen that every log-concave sequence is unimodal. More precisely, a log-concave sequence is either monotone (increasing or decreasing) or unimodal with one or two maximal values.

8.2.4. Conjectures. *Let L be a geometric lattice.*
(i) *The Whitney numbers $(w_k^+:0 \leqslant k \leqslant r)$ and $(W_k:0 \leqslant k \leqslant r)$ form unimodal sequences.*
(ii) *The Whitney numbers $(w_k^+:0 \leqslant k \leqslant r)$ and $(W_k:0 \leqslant k \leqslant r)$ form log-concave sequences.*

The remainder of this chapter will report on the progress towards proving these two conjectures of which (ii) is, of course, stronger than (i). While the general conjectures remain open, Conjecture 8.2.4 (ii) has been verified for several infinite classes of geometric lattices and there appears to be good hope that its truth may be established in the near future.

The main reason for mostly studying the stronger condition Conjecture 8.2.4 (ii) is that it is algebraically more readily accessible as the following Proposition 8.2.6 demonstrates. As a first result we note:

8.2.5. Proposition. *If the real polynomials $p_1(\lambda)$ and $p_2(\lambda)$ have log-concave coefficient sequences then so does the product $p_1(\lambda)p_2(\lambda)$.*

Proof. Just group the coefficients carefully together.   □

In view of Proposition 8.2.3 we may therefore confine ourselves to indecomposable geometric lattices when attempting to verify Conjecture 8.2.4 (ii). A very useful sufficient condition is given in the next result.

8.2.6. Proposition. *Let $p(\lambda) = \sum_{k=0}^{r} v_k \lambda^{r-k}$ be a polynomial with positive real coefficients v_k. If $p(\lambda)$ has only real roots then*

$$v_k^2 \geqslant v_{k-1}v_{k+1} \frac{k+1}{k} \frac{r-k+1}{r-k} \quad (1 \leqslant k \leqslant r-1).$$

Hence, the v_k's form a log-concave sequence with at most two maximal values.

Proof. Consider the polynomial $q(\lambda, \sigma) = \sum_{i=0}^{r} v_i \lambda^{r-i} \sigma^i$. Setting $\lambda = \sigma\tau$ we have $q(\lambda, \sigma) = \sum_{i=0}^{r} v_i \sigma^{r-i} \tau^{r-i} \sigma^i = \sigma^r \sum_{i=0}^{r} v_i \tau^{r-i} = \sigma^r p(\tau)$. Hence every root $(\lambda, \sigma) \neq (0,0)$ of q has real quotient $\tau = \lambda/\sigma$. Note that $\sigma \neq 0$ because $v_k > 0$. Applying Rolle's theorem we conclude that the same holds for $\partial q/\partial \lambda$ and $\partial q/\partial \sigma$ and hence by induction for every derivative $\partial^{i+j} q/\partial \lambda^i \partial \sigma^j$, and thus in particular for $\partial^{r-2} q/\partial \lambda^{r-k-1} \partial \sigma^{k-1}$. Substituting again $\lambda = \sigma\tau$ we obtain a polynomial of second degree of which the discriminant must be $\geqslant 0$, and this is precisely the inequality of the theorem.   □

8.2.7. Examples. Proposition 8.2.6 gives us the means to settle Conjecture 8.2.4 (ii) for the standard examples in Example 8.2.2. For the characteristic polynomial of the lattices B_n, $L(n, q)$, and Π_n this follows by a direct application of Proposition 8.2.6. Let us take a look at the Stirling numbers of the second kind. We set $\rho_n(\lambda) = p(\Pi_n; \lambda) = \sum_{k=1}^{n} S_{n,k} \lambda^{k-1}$. From the recursion formula for the numbers $S_{n,k}$ (Aigner 1979) we infer $\rho_{n+1}(\lambda) = \lambda\rho_n(\lambda) + \mathrm{d}(\lambda\rho_n(\lambda))/\mathrm{d}\lambda$. Setting $\beta_n(\lambda) = e^\lambda \rho_n(\lambda)$ we have $\beta_{n+1}(\lambda) = \lambda \mathrm{d}\beta_n(\lambda)/\mathrm{d}\lambda$ for all n. Hence $\beta_{n+1}(\lambda)$ has by induction only real roots and thus so does $\rho_n(\lambda)$, which proves the log-concavity for the sequence $(S_{n,k}: k = 1, \ldots, n)$. It is a still unproven conjecture that $(S_{n,k}: k = 1, \ldots, n)$ attains a unique maximum for $n \geqslant 3$.

We note further that, by Proposition 7.5.2 and Proposition 8.2.6, the sequence $(w_k^+ : k = 0, \ldots, r)$ is log-concave for any supersolvable lattice.

Whether this also holds (in the geometric case) for the sequence of Whitney numbers of the second kind is unknown.

8.3. The Möbius Algebra

To obtain sharper results on the Whitney numbers we must take a closer look at the behaviour of the Möbius function μ. Let us recall the definition (see Chapter 7). Let P be a finite poset. Then μ is an integer-valued function on the set $\mathrm{Int}(P)$ of non-empty intervals of P defined inductively by

$$\mu(a, a) = 1 \qquad\qquad (a \in P)$$

$$\mu(a, b) = - \sum_{a \leqslant x < b} \mu(a, x) = - \sum_{a < y \leqslant b} \mu(y, b) \quad (a < b). \qquad (8.6)$$

Alternatively, μ is the inverse of the zeta-function ζ in the incidence algebra of P. Having looked at it in this way we immediately deduce the *Möbius inversion principle*:

(i) Let f, g be functions on P into a field of characteristic 0. Then

$$g(a) = \sum_{x: x \leqslant a} f(x) \quad (a \in P) \Leftrightarrow f(a) = \sum_{x: x \leqslant a} \mu(x, a) g(x) \quad (a \in P)$$

Dually:

(ii) Let f, g be functions on P into a field of characteristic 0. Then

$$g(a) = \sum_{x: x \geqslant a} f(x) \quad (a \in P) \Leftrightarrow f(a) = \sum_{x: x \geqslant a} \mu(a, x) g(x) \quad (a \in P).$$

From now on, let L be a lattice. We denote by $V(L)$ the free vector space over $\mathbb{Q}$ generated by L where the basis element of $V(L)$ corresponding to $a \in L$ shall be denoted by ε_a. Hence the elements $\tau \in V(L)$ are all linear combinations $\tau = \sum_{a \in L} \tau(a)\varepsilon_a$ with $\tau(a)$ being the coefficient of ε_a.

8.3.1. Definition. The vector space $V(L)$ together with the componentwise product $\sum_{a \in L} \sigma(a)\varepsilon_a \cdot \sum_{a \in L} \tau(a)\varepsilon_a := \sum_{a \in L}(\sigma(a)\tau(a))\varepsilon_a$ is called the *Möbius algebra* of L (over $\mathbb{Q}$).

We define two sets ι_a $(a \in L)$ and $\kappa_a (a \in L)$ of elements in $V(L)$.

(i) $\iota_a := \sum_{x: x \leqslant a} \varepsilon_x \quad (a \in L),$

(ii) $\kappa_a := \sum_{x: x \vee a = 1} \varepsilon_x \quad (a \in L).$

(8.7)

8.3.2. Proposition. *Let $a \in L$. Then*:

(i) $\varepsilon_a = \sum_{x: x \leqslant a} \mu(x, a)\iota_x,$

(ii) $\kappa_a = \sum\limits_{x:x \geqslant a} \mu(x,1)\iota_x,$

(iii) $\mu(a,1)\iota_a = \sum\limits_{x:x \geqslant a} \mu(a,x)\kappa_x.$

Proof. Möbius inversion on part (i) of (8.7) yields (i). As to (ii), we have by (8.6) and (8.7)

$$\sum_{x:x \geqslant a} \mu(x,1)\iota_x = \sum_{x:x \geqslant a} \mu(x,1) \sum_{z:z \leqslant x} \varepsilon_z$$

$$= \sum_z \left\{ \sum_{x:x \geqslant z \vee a} \mu(x,1) \right\} \varepsilon_z = \sum_{z:z \vee a = 1} \varepsilon_z = \kappa_a.$$

(iii) now follows through Möbius inversion on (ii).   □

8.3.3. Corollary.
(i) *The set* $\{\iota_a : a \in L\}$ *is a basis of* $V(L)$.

(ii) *If* $\mu(a,1) \neq 0$ *for all* $a \in L$, *then* $\{\kappa_a : a \in L\}$ *is also a basis of* L, *and we have*

(iii) $\varepsilon_a = \sum\limits_{x \in L} v(a,x)\kappa_x$ *where* $v(a,x) = \sum\limits_{z:z \leqslant a \wedge x} \dfrac{\mu(z,a)\mu(z,x)}{\mu(z,1)}.$

From Proposition 8.3.2 (iii) we can now deduce a very useful formula for μ.

8.3.4. Corollary.
Let L *be a lattice and* $0 < b \in L$. *Then*

$$\sum_{x:x \vee b = 1} \mu(0,x) = 0.$$

Proof. Set $a = 0$ in Proposition 8.3.2 (iii) and compare coefficients for b.   □

Corollary 8.3.4, in turn, yields an easy proof of the fact that the Whitney numbers of the first kind of a geometric lattice have alternating sign.

8.3.5. Corollary.
Let L *be a geometric lattice of rank* r. *Then* $(-1)^r \mu(0,1) > 0$ *and thus* $w_k^+ = (-1)^k \sum\limits_{a:r(a)=k} \mu(0,a) > 0$ *for* $k = 0, 1, \ldots, r$. *As a consequence*, $\mu(a,b) \neq 0$ *whenever* $a \leqslant b$.

Proof. Since every interval of a geometric lattice is itself geometric we may assume by induction that $(-1)^{r-1}\mu(0,x) > 0$ for every copoint x. Let p be a point. Then by Corollary 8.3.4,

$$\mu(0,1) = - \sum_{\substack{x \text{ copoint} \\ p \nleqslant x}} \mu(0,x)$$

and thus $(-1)^r \mu(0,1) > 0$.   □

From the definition of the product in $V(L)$ we obtain another interesting description of μ. First, note that $\varepsilon_x \varepsilon_y = \delta_{xy}\varepsilon_x$ and thus $\iota_a \iota_b = \iota_{a \wedge b}$.

8.3.6. Proposition. *Let L be a lattice with copoint-set C. Then $\mu(0,1) = \sum_{k\leqslant 0} (-1)^k r_k$ where r_k is the number of k-sets $A \subseteq C$ with $\inf A = 0$. In particular, $\mu(0,1) = 0$ if 0 is not a meet of copoints.*

Proof. Let $\iota_c \in V(L)$ where $c \in C$. Then clearly $\iota_1 - \iota_c = \sum_{x:x \not\leqslant c} \varepsilon_x$, and thus $\varepsilon_1 = \prod_{c \in C}(\iota_1 - \iota_c)$ since ε_1 is the only ε_x which appears in all factors of the right-hand side. If we now express both sides in terms of the basis $\{\iota_x : x \in L\}$ and compare coefficients for $0 \in L$, the formula results. $\square$

The expression for $\mu(0,1)$ just obtained furnishes an alternative description for the characteristic polynomials of geometric lattices, which turns out to be useful in the study of Whitney numbers.

8.3.7. Corollary. *Let L be a geometric lattice with point-set E. Then $p(L;\lambda) = \sum_{A:A\subseteq E}(-1)^{|A|}\lambda^{r(1)-r(\sup A)}$.*

Proof. This is a restatement of Proposition 7.2.1. $\square$

Let us return to Corollary 8.3.3. The fact that the κ_a's form a basis of $V(L)$ can be translated into a very interesting property of the lattice L.

8.3.8. Proposition. *Suppose L is a lattice and $\mu(a,1) \neq 0$ for all $a \in L$. Then there exists a bijection $\phi : L \to L$ such that $a \vee \phi(a) = 1$ for all a.*

Proof. By Corollary 8.3.3(ii), the κ_a's form a basis. Writing them in terms of the basis $\{\varepsilon_a : a \in L\}$, we conclude that the integral matrix $[k_{ab}]_{a,b \in L}$ with

$$k_{ab} = \begin{cases} 1 & \text{if } a \vee b = 1, \\ 0 & \text{otherwise} \end{cases}$$

is non-singular. Hence some term in the determinant expansion does not vanish which is precisely the statement of the theorem. $\square$

The dual statement holds, of course, as well and we may combine the two to give:

8.3.9. Proposition. *Let L be a lattice such that $\mu(0,a)\mu(a,1) \neq 0$ for all $a \in L$. Then there exists a bijection $\phi : L \to L$ with $a \vee \phi(a) = 1$, $a \wedge \phi(a) = 0$ for all $a \in L$.*

Proof. We define the elements $\tau_a \in V(L), a \in L$, by $\tau_a = \sum_{x:x \leqslant a}\mu(0,x)\kappa_x$. By Möbius inversion we have $\mu(0,a)\kappa_a = \sum_{x:x \leqslant a}\mu(x,a)\tau_x$ for $a \in L$, hence $\{\tau_a : a \in L\}$ is also a basis of $V(L)$. Using (8.7)(ii) we have $\tau_a = \sum_{x:x \leqslant a}\mu(0,x)\sum_{z:z \vee x = 1}\varepsilon_z = \sum_z[\sum_{x:x \leqslant a,x \vee z = 1}\mu(0,x)]\varepsilon_z$. If $a \vee z < 1$, then the coefficient of ε_z is 0, so assume $a \vee z = 1$. In order to imitate the argument in the proof of Proposition 8.3.8 it remains to be shown that $\sum_{x:x \leqslant a,x \vee z = 1}\mu(0,x) = 0$ whenever $a \wedge z > 0$. This is certainly true for $z = 1$, so assume $z < 1$ and let $Q = \{x : x = a$ or

$x < a$, $x \vee z < 1\} \subseteq [0, a]$ with Möbius function μ_Q. Since x, $y \in Q$ implies $x \wedge y \in Q$, Q is a lattice. We define the mapping $x \mapsto \bar{x}$ from $[0, a]$ to Q by

$$\bar{x} = \begin{cases} x & \text{if } x \vee z < 1 \\ a & \text{if } x \vee z = 1. \end{cases}$$

By our hypothesis, $\bar{0} = 0$. Now we have

$$\sum_{x:x \leqslant a, x \vee z = 1} \mu(0, x) = \sum_{x:x \leqslant a, \bar{x} = a} \mu(0, x) = \sum_{x:x \leqslant a} \mu(0, x)\delta_Q(\bar{x}, a)$$

$$= \sum_{x:x \leqslant a} \mu(0, x) \sum_{\bar{w} \in Q: \bar{x} \leqslant \bar{w}} \mu_Q(\bar{w}, a)$$

$$= \sum_{x:x \leqslant a} \mu(0, x) \sum_{\bar{w} \in Q: x \leqslant \bar{w}} \mu_Q(\bar{w}, a)$$

$$\text{(since} \quad \bar{x} \leqslant \bar{w} \Leftrightarrow x \leqslant \bar{w})$$

$$= \sum_{\bar{w} \in Q} \left\{ \sum_{x:x \leqslant \bar{w}} \mu(0, x) \right\} \mu_Q(\bar{w}, a) = \mu_Q(0, a).$$

Now it is easily seen that $a \wedge z$ is a lower bound for all copoints of Q whence by Proposition 8.3.6 we conclude $\mu_Q(0, a) = 0$ whenever $a \wedge z > 0$.   □

8.3.10. Corollary. *In a geometric lattice L there exists a bijection $\phi : L \to L$ such that $a \vee \phi(a) = 1$, $a \wedge \phi(a) = 0$ for all $a \in L$.*

8.4. The Whitney Numbers of the First Kind

Let L be a geometric lattice of rank r. By Definition 8.2.1 and Corollary 8.3.5, the polynomial $\psi(L:\lambda) = \sum_{i=0}^{r} w_i^+ \lambda^{r-i} = (-1)^r p(L; -\lambda)$ has positive coefficients w_i^+. Using Corollary 8.3.7, we may rewrite $\psi(L; \lambda)$ as

$$\psi(L; \lambda) = \sum_{A:A \subseteq E} (-1)^{|A| - r(\sup A)} \lambda^{r - r(\sup A)}. \tag{8.8}$$

The right-hand side of (8.8) is, of course, the value of the rank generating function evaluated at $(\lambda, -1)$ (see Crapo 1969, Heron 1972, and Chapter 7). Equation (8.8) permits an inductive argument which establishes the following major result.

8.4.1. Theorem. *Let L be a geometric lattice of rank r. Then*

$$\psi(L; \lambda) = \sum_{i=0}^{r} t_i (\lambda + 1)^{r-i}$$

with all coefficients $t_i \in \mathbb{Z}$, $t_i \geqslant 0$.

Proof. For the lattice of rank 1 we have, by (8.8), $\psi(L;\lambda) = \lambda + 1$. Suppose the theorem holds for all geometric lattices of rank at most $r - 1$, and let L have rank r. Within rank r we use induction on the cardinality $|E|$ of points in L. If $|E| = r$ then $L = B_r$ is the Boolean algebra. Hence, by Examples 8.2.2, $\psi(L;\lambda) = (\lambda + 1)^r$. Otherwise, there exists a point p not in the center of L (i.e., $[0, p]$ is not a factor of L) and by separating the sets $A \subseteq E$ according as to whether or not they contain p, we obtain

$$\psi(L;\lambda) = \psi(L';\lambda) + \psi(L'';\lambda) \tag{8.9}$$

where L' and L'' are the lattices of flats of the restriction to $E - p$ and contraction through p, respectively. Now, L' has one point less than L and $L' \cong [p, 1]$ has rank one less than L. Thus by induction their polynomials ψ have the form as required in the theorem and hence so does $\psi(L;\lambda)$. $\square$

Theorem 8.4.1 allows us to deduce the result that the unimodality of the (unsigned) Whitney numbers of the first kind can only fail to hold in the upper half of the lattice.

8.4.2. Corollary. *Let L be a geometric lattice of rank $r \geqslant 2$, and suppose $L \neq B_r$. Then*

   (i) $w_k^+ < w_l^+$ *for $1 \leqslant k < r/2$ and $k < l \leqslant r - k$.*
In particular,
   (ii) $w_0^+ < w_1^+ < \cdots < w_{[(r+1)/2]}^+$,
   (iii) $w_k^+ < w_{r-k}^+$ *for $0 \leqslant k < r/2$.*

Proof. By Theorem 8.4.1,

$$w_k^+ = \sum_{i=0}^{k} t_i \binom{r-i}{k-i}, \qquad t_k = \sum_{i=0}^{k} (-1)^{k-i} w_i^+ \binom{r-i}{k-i}. \tag{8.10}$$

Hence $t_0 = 1, t_1 = w_1^+ - r = W_1 - r > 0$ (since $L \neq B_r$), and $t_j \geqslant 0$ for $j \geqslant 2$. Now, for $0 \leqslant k < r/2$ and $k < l \leqslant r - k$ we have $\binom{r-i}{k-i} \leqslant \binom{r-i}{l-i}$ for $i = 0, \ldots, k$, and therefore $t_i \binom{r-i}{k-i} \leqslant t_i \binom{r-i}{l-i}$ with inequality holding for $i = 1$. $\square$

It is now apparent that further knowledge of the coefficients t_i will imply stronger conditions on the sequence $(w_k^+ : k = 0, \ldots, r)$. (See Björner 1987, Brylawski 1977a, and Heron 1972.) The formulae (8.10) give as a byproduct an interesting lower bound for the w_k^+'s.

8.4.3. Corollary. *Let L be a geometric lattice of rank $r \geqslant 2$ and point-set E, $|E| = n$. Suppose that all e-subsets of E are independent (i.e., $r(\sup A) = |A|$ for*

$|A| \leqslant e$). *Then*

$$w_k^+ \geqslant \sum_{i=0}^{e-1} \binom{n-r+i-1}{i} \binom{r-i}{k-i} \quad (0 \leqslant k \leqslant r).$$

In particular,

$$|\mu(0,1)| \geqslant \sum_{i=0}^{e-1} \binom{n-r+i-1}{i}.$$

Proof. By (8.10) it suffices to show that $t_i = \binom{n-r+i-1}{i}$ for $0 \leqslant i \leqslant e-1$.

By the hypothesis, $[0,a] \cong B_i$, if a has rank $i \leqslant e-1$. Thus $W_i = w_i^+ = \binom{n}{i}$

for $i = 0,\dots,e-1$. We now conclude from (8.10) for $k \leqslant e-1$:

$$t_k = \sum_{i=0}^{k} (-1)^{k-i} \binom{r-i}{k-i}\binom{n}{i}$$

$$= \sum_{i=0}^{k} \binom{-r+k-1}{k-i}\binom{n}{i}$$

$$= \binom{n-r+k-1}{k}. \qquad \qquad \square$$

Since in a geometric lattice all 2-subsets are independent we may supplement Corollary 8.4.3 by listing two general bounds for the Whitney numbers w_k^+ of any geometric lattice.

8.4.4. Corollary. *Let L be a geometric lattice of rank $r \geqslant 2$ and with n points. Then*

$$\binom{r}{k} + (n-r)\binom{r-1}{k-1} \leqslant w_k^+ \leqslant \binom{n}{k} \quad (k = 0,\dots,r).$$

The only lattices that attain equality on the upper bound for some $k \geqslant 2$ (or equivalently all k) are the Boolean algebras B_r. The only lattices that attain equality on the lower bound for some $k \geqslant 2$ (or equivalently all k) are direct products of a rank 2 lattice with a Boolean algebra.

Proof. The upper bound for w_k^+ and the characterization of the extremal lattices as B_r follow directly from the inductive argument in (8.9). The lower bound is just the case $e = 2$ in Corollary 8.4.3. Suppose now the lower bound is attain for some $k \geqslant 2$. By (8.10) this implies $t_i = 0$ for $2 \leqslant i \leqslant k$, and thus

$$w_i^+ = \binom{r}{i} + (n-r)\binom{r-1}{i-1} \text{ for } 0 \leqslant i \leqslant k. \text{ We show next that if } w_k^+ (k \geqslant 2)$$

attains the lower bound then so does w_{k+1}^+. If $L = B_r$, then there is nothing to show. Otherwise, there exists a point not in the center and we may apply (8.9) which says

$$w_k^+ = w_k' + w_{k-1}'' \qquad (8.11)$$

where w_i', w_i'' are the (unsigned) i-th Whitney numbers of the lattices L' and L'', respectively. Hence

$$\binom{r}{k} + (n-r)\binom{r-1}{k-1} = w_k^+$$

$$= w_k' + w_{k-1}'' \geqslant \binom{r}{k} + (n-1-r)\binom{r-1}{k-1}$$

$$+ \binom{r-1}{k-1} + (m-r+1)\binom{r-2}{k-2}$$

$$= \binom{r}{k} + (n-r)\binom{r-1}{k-1} + (m-r+1)\binom{r-2}{k-2} \qquad (8.12)$$

where $m \leqslant n-1$ is the number of points in L''.

Hence w_k', w_{k-1}'' both attain the lower bound and further $m = r - 1$. By induction, we conclude that w_{k+1}', w_k'' attain the lower bound as well. Using (8.11) with $k+1$ instead of k the same must then be true for w_{k+1}^+. Since $L = L_1 \times L_2$ implies $\psi(L;\lambda) = \psi(L_1;\lambda)\psi(L_2;\lambda)$, the lattices mentioned in the theorem attain the lower bounds. Suppose now L attains the lower bound for all k, where $L \neq B_r$. Then from (8.12) L' and L'' also attain the lower bounds, and in addition $m = r - 1$, i.e., $L'' = [p, 1] \cong B_{r-1}$. By induction, $L' \cong B_{r-2} \times M_2$ and $r(M_2) = 2$, which together with $L'' \cong B_{r-1}$ yields the theorem. $\qquad\square$

For special classes of lattices the bounds in Corollary 8.4.4 can be considerably sharpened, e.g., for indecomposable lattices (see Björner 1987, Brylawski 1977a).

8.4.5. Example. In Section 7.5 the chromatic polynomial $\chi(\Gamma;\lambda)$ of a graph Γ was related to the characteristic polynomial $p(L;\lambda)$ of the lattice L of the flats of the graphic matroid associated with Γ by means of

$$\chi(\Gamma;\lambda) = \lambda^{c(\Gamma)} p(L;\lambda),$$

where $c(\Gamma)$ is the number of connected components of Γ. Hence the results of this section, in particular Corollary 8.4.2, apply to the coefficients of $\chi(\Gamma;\lambda)$. Whether the absolute values of the coefficients do indeed form a unimodal sequence is still open.

8.5. The Whitney Numbers of the Second Kind

We have seen in Section 8.2 that the numbers W_k form a unimodal sequence, in fact, a log-concave sequence for the standard examples B_n, $L(n, q)$, Π_n. A few more classes are known to possess this property, e.g. affine lattices, Hartmanis (partition) lattices (see Aigner 1979, p. 258), the lattices of matroid designs (see Young, Murty, and Edmonds 1970), and, in general, all geometric lattices with up to eight points (see Blackburn, Crapo, and Higgs 1973).

In analogy to the Whitney numbers of the first kind, as a first step towards verifying Conjecture 8.2.4(i) one would like to prove Corollary 8.4.2 for the numbers W_k. While this has not yet been done, the following result points in this direction.

8.5.1. Proposition. *Let L be a geometric lattice of rank r. Then*

$$W_1 + W_2 + \cdots + W_k \leqslant W_{r-1} + W_{r-2} + \cdots + W_{r-k} \quad (1 \leqslant k \leqslant r-1), \qquad (8.13)$$

with equality for some k if and only if L is modular.

Proof. Let $V(L)$ be the free vector space of L over $\mathbb{Q}$ as defined in Section 8.3 and denote by $V_k(L)$ the subspace generated by $\{\varepsilon_a : a \in L, r(a) \leqslant k\}$. Thus, dim $V_k(L) = \sum_{i=0}^{k} W_i$. Let $\pi : V(L) \to V_k(L)$ be the linear projection onto $V_k(L)$, i.e.,

$$\pi(\varepsilon_a) = \begin{cases} \varepsilon_a & \text{if} \quad r(a) \leqslant k \\ 0 & \text{otherwise.} \end{cases}$$

From the definition (8.7)(ii) and the semimodular rank inequality we infer $\pi(\kappa_b) = 0$ whenever $r(b) < r - k$. Since the set $\{\kappa_b : b \in L\}$ is a basis of $V(L)$ we infer from Corollary 8.3.3 (ii) that the set $\{\pi(\kappa_b) : r(b) \geqslant r - k\}$ spans $V_k(L)$ and hence that the stated inequality holds. For modular lattices we have equality since the dual of a modular geometric lattice is again geometric. Now suppose we have equality in (8.13) for some k. By our argument, this implies that $\{\pi(\kappa_b) : r(b) \geqslant r - k\}$ is a basis of $V_k(L)$ and, in particular, linearly independent. Take $a \in L$ with $r(a) > k$. Then by Corollary 8.3.3 (iii), we have $0 = \pi(\varepsilon_a) = \sum_{x:r(x) \geqslant r-k} v(a,x)\pi(\kappa_x)$, and hence $v(a,x) = \sum_{z:z \leqslant a \wedge x} \mu(z,a)\mu(z,x)/\mu(z,1) = 0$ for all a, $x \in L$ with $r(a) \geqslant k+1$ and $r(x) \geqslant r-k$. This implies, in particular, that $a \wedge x > 0$ for all such pairs since otherwise $v(a,x) = \mu(0,a)\mu(0,x)/\mu(0,1) \neq 0$. But this last condition, as is well-known, implies the modularity of L. $\square$

The k-truncation $L^{(k)}$ of L is obtained by identifying all elements of L with rank $\geqslant k$, keeping the lower part up to rank $k-1$ unchanged. If L is geometric then, clearly, so is $L^{(k)}$. Hence we have:

8.5.2. Corollary. *Let L be a geometric lattice of rank $r \geqslant 3$. Then*

(i) $\displaystyle\sum_{i=1}^{k} W_i < \sum_{i=1}^{k} W_{l-i} \quad (1 \leqslant k \leqslant l-2, l \leqslant r-1)$,

(ii) $\displaystyle\sum_{i=1}^{k} W_i \leqslant \sum_{i=1}^{k} W_{r-i}$, with equality for some k iff L is modular.

Proof. We have just seen (ii). (i) now follows easily by considering the truncation $L^{(l)}$. $\square$

Let us denote by $\text{Bot}_k(L)$ and $\text{Top}_k(L)$ the lower part of L from rank 1 up to rank k and the top part from corank 1 down to corank k, i.e., $\text{Bot}_k(L) = \{a \in L: 1 \leqslant r(a) \leqslant k\}$, $\text{Top}_k(L) = \{a \in L: r - k \leqslant r(a) \leqslant r - 1\}$. Thus $|\text{Bot}_k(L)| = \sum_{i=1}^{k} W_i$, $|\text{Top}_k(L)| = \sum_{i=1}^{k} W_{r-i}$. Proposition 8.5.1 raises the question whether there exist injections from $\text{Bot}_k(L)$ into $\text{Top}_k(L)$ using the ordering relation of L or its complement.

8.5.3. Proposition. *Let L be a geometric lattice of rank $r \geqslant 2$ and let $1 \leqslant k \leqslant r - 1$. Then there exist injections f, g:*

(i) *$f: \text{Bot}_k(L) \to \text{Top}_k(L)$ with $a \leqslant f(a)$ for all $a \in \text{Bot}_k(L)$,*

(ii) *$g: \text{Bot}_k(L) \to \text{Top}_k(L)$ with $a \nleqslant g(a)$ for all $a \in \text{Bot}_k(L)$.*

Proof. Proposition 8.3.9 proves (ii) by noticing that any complementing map takes 0 into 1. Let $\pi: V(L) \to V_k(L)$ be the projection of Proposition 8.5.1. We observed in the proof of Proposition 8.5.1 that $\{\pi(\kappa_a): r(a) \geqslant r - k\}$ spans $V_k(L)$. Hence, by Proposition 8.3.2 (ii), $\{\pi(\iota_a): r(a) \geqslant r - k\}$ also spans $V_k(L)$. Therefore, the matrix $I = (i_{a,b})$ indexed by $\{a \in L: r(a) \leqslant k\}$ and $\{b \in L: r(b) \geqslant r - k\}$ with

$$i_{a,b} = \begin{cases} 1 & \text{if } a \leqslant b \\ 0 & \text{if } a \nleqslant b \end{cases}$$

has rank $\sum_{i=0}^{k} W_i$ which implies the existence of an injection $f: \text{Bot}_k(L) \cup \{0\} \to \text{Top}_k(L) \cup \{1\}$ with $a \leqslant f(a)$ by the same argument as in the proof of Proposition 8.3.8. It remains to show that f can be chosen to map 0 into 1. To this end, it suffices, because of $i_{0,1} = 1$, to prove that the submatrix $I' \subseteq I$ consisting of all rows except row 0 and all columns except column 1 has rank $\sum_{i=1}^{k} W_i$. Now, by Corollary 8.3.3 (iii) and Proposition 8.3.2 (ii),

$$\varepsilon_0 = \sum_{x: r(x) \geqslant r - k} v(0, x) \pi(\kappa_x)$$

$$= \sum_{x: r(x) \geqslant r - k} v(0, x) \sum_{w: w \geqslant x} \mu(w, 1) \pi(\iota_w)$$

$$= \sum_w \left(\sum_{\substack{x: r(x) \geqslant r - k \\ x \leqslant w}} v(0, x) \mu(w, 1) \right) \pi(\iota_w).$$

The coefficient of $\pi(\iota_1)$ in the last sum is

$$\frac{1}{\mu(0, 1)} \sum_{x: r(x) \geqslant r - k} \mu(0, x) = \frac{1}{\mu(0, 1)} \left(- \sum_{x: r(x) < r - k} \mu(0, x) \right)$$

$$= \frac{1}{\mu(0, 1)} \mu_{L^{(r-k)}}(0, 1) \neq 0.$$

We conclude that $\pi(\iota_1)$ is in the span of $\{\pi(\iota_a): r - k \leqslant r(a) < r\} \cup \varepsilon_0$, and hence that this latter set also spans $V_k(L)$. There exists therefore a non-singular square submatrix of $(I - \text{column } 1) \cup \text{column } 0'$ of rank $\sum_{i=0}^{k} W_i$, where $0'$ has a one in row 0 and zeros elsewhere, and thus a required non-singular square submatrix of I' of rank $\sum_{i=1}^{k} W_i$.

The inequalities in Proposition 8.5.1 can be sharpened, as spelled out in the following result.

8.5.4. Proposition. *Let L be a geometric lattice of rank $r \geqslant 2$, let $1 \leqslant k \leqslant r - 1$, and let $a \in L$ with $r(a) \leqslant r - k$. Then*

$$|Bot_k(L)| \leqslant |Top_k(L)| - |Top_k([a, 1])|$$
$$+ |Bot_k([a, 1])| \leqslant |Top_k(L)|.$$

Proof. The inequality on the right is clear by Proposition 8.5.1. Let $\alpha_z \in V(L)$, $z \in L$, be defined by $\alpha_z = \sum_{x : x \vee a = z} \varepsilon_x$. Then $\iota_b = \sum_{z : a \leqslant z \leqslant b} \alpha_z$ for all $b \in L$, $b \geqslant a$. If $\pi : V(L) \to V_k(L)$ is the projection as before, it follows from the semimodular inequality that $\pi(\alpha_z) = 0$ for all $z \in L, z \geqslant a$ with $r(z) > r(a) + k$. Since we have seen in the proof of Proposition 8.5.3 that $\{\pi(\iota_b) : r(b) \geqslant r - k\}$ spans $V_k(L)$ it follows that the set $\{\pi(\iota_b) : b \not\geqslant a, r(b) \geqslant r - k\} \cup \{\pi(\alpha_z) : z \geqslant a, r(z) \leqslant r(a) + k\}$ also spans $V_k(L)$, and this is just the required inequality. $\qquad \square$

Note that Proposition 8.5.4 implies again that $|Bot_k(L)| = |Top_k(L)|$ forces the lattice to be modular, since it is known (see Wille 1971) that a geometric lattice L of rank $r \geqslant 5$ is modular if all of its intervals $[p, 1]$, p point, are modular (the cases $r \leqslant 4$ being trivial).

In analogy to Corollary 8.4.4 let us now bound the Whitney numbers of the second kind for arbitrary geometric lattices.

8.5.5. Proposition. *Let L be a geometric lattice of rank $r \geqslant 3$ and with n points. Then*

$$\binom{r-2}{k-1}(n-r) + \binom{r}{k} \leqslant W_k \leqslant \binom{n}{k} \qquad (0 \leqslant k \leqslant r).$$

Equality holds on the right for all k iff $L = B_r$, and on the left for all k iff L is a direct product of a modular lattice of rank 3 and a Boolean algebra.

Proof. The inequality on the right is trivial since any k independent points span a unique flat of rank k. That equality for all k characterises B_r is also clear. For the first inequality we use induction on the rank r. For $r = 3$ the result is true by Proposition 8.5.1 including the case of equality. Within rank r we use induction on k. Again, for $k = 0$ or 1 there is nothing to prove. So let us assume $r \geqslant 4$ and $k \geqslant 2$ where by Proposition 8.5.1 we may assume $k \leqslant r - 2$. The following notation is useful:

$$A_j(u, v) = \{x \in [u, v] : r(x) = j\}, \quad a_j(u, v) = |A_j(u, v)|,$$
$$B_j(u, v) = \{x \notin [u, v] : r(x) = j\}, \quad b_j(u, v) = |B_j(u, v)|.$$

Let p be a point of L. Then $W_k = a_k(p, 1) + b_k(p, 1)$. If p is covered by l lines then, by induction,

$$a_k(p, 1) \geqslant \binom{r-3}{k-2}(l-r+1) + \binom{r-1}{k-1}. \tag{8.14}$$

The mapping $x \mapsto x \vee p$ from $B_k(p, 1)$ to $A_{k+1}(p, 1)$ is a surjection mapping $a_k(0, y) - a_k(p, y)$ elements of $B_k(p, 1)$ onto y for each $y \in A_{k+1}(p, 1)$. Now, by Proposition 8.5.4 applied to $[0, y]$, we have $a_k(0, y) - a_k(p, y) \geqslant a_1(0, y) - a_2(p, y)$ and thus

$$b_k(p, 1) \geqslant \sum_{y \in A_{k+1}(p,1)} a_1(0, y) - \sum_{y \in A_{k+1}(p,1)} a_2(p, y).$$

Interchanging the order of summation in each of the summands we obtain

$$\sum_{y \in A_{k+1}(p,1)} a_1(0, y) = a_{k+1}(p, 1) + \sum_{g \in A_2(p,1)} (a_1(0, g) - 1)a_{k+1}(g, 1)$$

and

$$\sum_{y \in A_{k+1}(p,1)} a_2(p, y) = \sum_{g \in A_2(p,1)} a_{k+1}(g, 1).$$

Using induction and the fact that $\sum_{g \in A_2(p,1)} (a_1(0, g) - 1) = n - 1$, we infer

$$b_k(p, 1) \geqslant \binom{r-3}{k-1}(l-r+1) + \binom{r-1}{k} + (n-1-l)\binom{r-2}{k-1} \tag{8.15}$$

and thus by (8.14) and (8.15)

$$W_k = a_k(p, 1) + b_k(p, 1) \geqslant \binom{r-2}{k-1}(n-r) + \binom{r}{k}.$$

As for equality, it is easy to verify that any direct product of a modular lattice of rank 3 and a Boolean algebra does indeed satisfy the left-hand side with equality. Suppose now, L satisfies the left-hand side with equality for all k. Then by Proposition 8.5.1 L is modular and if there is no line with more than two points, L must be a Boolean algebra. So we may suppose that there is a line g containing at least three points. Since we must have equality in (8.15), it follows that $[g, 1] \cong B_{r-2}$. Hence if h is a minimal complement of g then $[0, h] \cong B_{r-2}$ (cf. Aigner 1979, Proposition 2.4.4). If for every point p either $p \leqslant g$ or $p \leqslant h$, we have $L \cong [0, g] \times B_{r-2}$ and so we are finished (see Aigner 1979, Theorem 2.45). Hence let p be a point with $p \nleqslant g$ and $p \nleqslant h$. Let c be the modular plane $g \vee p$. Then there is a unique point $p' \leqslant h$ with $c = g \vee p'$ (remember $[g, 1] \cong B_{r-2}$). In order to show $L \cong [0, c] \times B_{r-3}$ we have to prove that for every point $q, q \nleqslant g, q \nleqslant h$, we must have $q \leqslant c$. Suppose not. Then let $q' \leqslant h$ be the unique point with $g \vee q = g \vee q'$. By induction, we must have $[q', 1] \cong [q', q' \vee c] \times [q', h']$ where $[q', h'] \cong B_{r-4}$ and h' is spanned by

$\{q' \vee s : s \leqslant h,\ s \neq q', p'\}$. But now, clearly, $q \vee q' \nleqslant q' \vee c$ and $q \vee q' \nleqslant h'$, a contradiction. $\square$

The striking similarity of Proposition 8.5.5 and Corollary 8.4.4 raises the question whether the Whitney numbers of the second kind are the face numbers of some (shellable) complex, the same way the Whitney numbers of the first kind are the face numbers of the broken circuit complex (see Brylawski 1977b). If so, inequalities analogous to Corollary 8.4.2 could be deduced for the numbers W_k as well.

As in the case of the Whitney numbers of the first kind, additional hypotheses on the structure of L allow sharper bounds in Proposition 8.5.5.

Proposition 8.5.3 (i) says for $k = 1$ that we can always map $L_1 = \{a \in L : r(a) = 1\}$ injectively into $L_l = \{a \in L : r(a) = l\}$, $1 \leqslant l \leqslant r - 1$, respecting the order relation in L. That this matching property does not hold in general between any two consecutive levels L_k and L_{k+1} was shown in Dilworth and Greene (1971). It is, surprisingly, not even satisfied in the lattice of partitions (see Canfield 1978). There is wide belief, however, that we can always match L_k into L_{r-k}. That there is always a partial matching of cardinality W_1 between any two consecutive levels is shown in the following result which is established by an application of Menger's theorem to L, viewed as a directed graph from 0 to 1.

8.5.6. Proposition. *In a geometric lattice L there exist W_1 maximal pairwise disjoint chains from the points into the copoints of L.*

Let us, finally, consider the log-concave property

$$W_k^2 \geqslant W_{k-1}W_{k+1} \quad (1 \leqslant k \leqslant r - 1) \tag{8.16}$$

for the sequence $(W_0, W_1, \ldots, W_r)$ of a geometric lattice L.

Inequality (8.16) is true for $k = 1$, since $W_2 \leqslant \binom{W_1}{2} < W_1^2$, and for $k = r - 1$, since any coline is the infimum of a pair of copoints whence $W_{r-2} \leqslant \binom{W_{r-1}}{2} < W_{r-1}^2$. By considering the truncation $L^{(k+2)}$ it is clear that we may assume $r = k + 2$ when proving (8.16).

Since (8.16) implies $W_k/W_{k+l} \leqslant W_{k+j}/W_{k+j+l}$ for all k, j and l, it may be easier to establish such 'wider' inequalities. Let again L_k denote the set of elements of rank k. Inequality (8.16) suggests looking for a suitable injection from $L_{k-1} \times L_{k+1}$ into $L_k \times L_k$.

8.5.7. Proposition. *Let L be a binary geometric lattice of rank r and let $2 \leqslant k \leqslant r$. Then $W_1 W_k \leqslant W_2 W_{k-1}$; in particular, $W_1 W_3 \leqslant W_2^2$.*

Proof. Assume first L to have rank $k + 1$. To L we associate a bipartite graph

G with vertex sets

$$A = \{(p, H_k) \in L_1 \times L_k : p \vee H_k = 1\} \quad \text{and}$$
$$B = \{(I_2, J_{k-1}) \in L_2 \times L_{k-1} : I_2 \vee J_{k-1} = 1\}$$

where (p, H_k) and (I_2, J_{k-1}) are joined by an edge iff there exists a point q such that $H_k = q \vee J_{k-1}$ and $I_2 = q \vee p$. For $A_1 \subseteq A$ let $b(A_1) \subseteq B$ be the set of vertices in B joined to some vertex in A_1, and similarly $a(B_1) \subseteq A$ for $B_1 \subseteq B$. For each $F_3 \in L_3$ we define analogously a bipartite graph $G(F_3)$ with vertex sets $A(F_3) = \{(p, H_2) \in L_1 \times L_2 : p < F_3, H_2 < F_3, \quad p \vee H_2 = F_3\}$ and $B(F_3) = \{(I_2, r) \in L_2 \times L_1 : I_2 < F_3, r < F_3, I_2 \vee r = F_3\}$ where (p, H_2) and (I_2, r) are joined by an edge iff there exists a point q such that $H_2 = q \vee r$ and $I_2 = q \vee p$. Obviously, $|A(F_3)| = |B(F_3)|$ and it is easily checked that $G(F_3)$ always contains a perfect matching, i.e., a matching containing $|A(F_3)|$ edges. (There are only six binary lattices of rank 3.)

Now pick $H_{k-2} \in L_{k-2}$ such that $F_3 \vee H_{k-2} = 1$. By mapping each $(p, H_2) \in A(F_3)$ onto $(p, H_2 \vee H_{k-2}) \in A$, each $(I_2, r) \in B(F_3)$ onto $(I_2, r \vee H_{k-2}) \in B$, and joining $(p, H_2 \vee H_{k-2})$ and $(I_2, r \vee H_{k-2})$ iff (p, H_2) and (I_2, r) are joined in $G(F_3)$, we obtain a subgraph $G(F_3, H_{k-2})$ of G. Since clearly $|A(F_3)| = |A(F_3, H_{k-2})|$ and $|B(F_3)| = |B(F_3, H_{k-2})|$, the two graphs $G(F_3)$ and $G(F_3, H_{k-2})$ are, in fact, isomorphic for each H_{k-2}. As is easily seen,

$$a(B(F_3, H_{k-2})) = A(F_3, H_{k-2}) \quad \text{and}$$
$$b(A(F_3, H_{k-2})) \supseteq B(F_3, H_{k-2}). \tag{8.17}$$

Since any $(p, H_k) \in A$ can be written as $(p, H_2 \vee H_{k-2})$ and similarly any $(I_2, J_{k-1}) \in B$ as $(I_2, r \vee H'_{k-2})$ for some H_2, H_{k-2}, r and H'_{k-2}, we have

$$A = \bigcup_{\substack{(F_3, H_{k-2}) \in L_3 \times L_{k-2} \\ F_3 \vee H_{k-2} = 1}} A(F_3, H_{k-2}) \quad \text{and} \quad B = \bigcup_{\substack{(F_3, H_{k-2}) \in L_3 \times L_{k-2} \\ F_3 \vee H_{k-2} = 1}} B(F_3, H_{k-2}).$$

Listing the subgraphs $G(F_3, H_{k-2})$ in some linear order we may [by (8.17)] extend the matchings step by step until all of A is exhausted whence $|A| \leqslant |B|$.

Now let L have arbitrary rank $r \geqslant k + 1$. Then we may write

$$W_1 W_k = \left| \bigcup_{G_{k+1} \in L_{k+1}} \{(G_1, G_k) \in L_1 \times L_k : G_1 \vee G_k = G_{k+1}\} \right|$$

$$+ \left| \bigcup_{G_k \in L_k} \{(G_1, G_k) \in L_1 \times L_k : G_1 < G_k\} \right|$$

and

$$W_2 W_{k-1} = \left| \bigcup_{G_{k+1} \in L_{k+1}} \{(G_2, G_{k-1}) \in L_2 \times L_{k-1} : G_2 \vee G_{k-1} = G_{k+1}\} \right|$$

$$+ \left| \bigcup_{G_k \in L_k}^{\cdot} \{(G_2, G_{k-1}) \in L_2 \times L_{k-1} : G_2 \vee G_{k-1} = G_k\} \right|$$

$$+ \left| \bigcup_{G_{k-1} \in L_{k-1}}^{\cdot} \{(G_2, G_{k-1}) \in L_2 \times L_{k-1} : G_2 \leqslant G_{k-1}\} \right|.$$

We have just shown that the first summand in $W_1 W_k$ is (for each G_{k+1}) less than or equal to the first summand in $W_2 W_{k-1}$ (for the corresponding G_{k+1}). The same holds for each $G_k \in L_k$ in the second summands since in $W_1 W_k$ the number of points below G_k are counted whereas in $W_2 W_{k-1}$ each copoint below G_k is counted at least once.

Unfortunately, the argument used in Proposition 8.5.7 cannot be generalized to arbitrary geometric lattices since not every geometric lattice of rank 3 has a matching as required in Proposition 8.5.7. But the following result certainly adds strength to the log-concave conjecture (8.16). To verify (8.16) for k it suffices to consider lattices L of rank $k + 2$ as remarked before Proposition 8.5.7. Now by the same argument as in the proof of Proposition 8.5.7 it would be enough to prove

$$|L| \leqslant |R| \tag{8.18}$$

where

$$L = |\{(G_{k-1}, G_{k+1}) \in L_{k-1} \times L_{k+1}, G_{k-1} \vee G_{k+1} = 1\}|$$

and

$$R = |\{(I_k, J_k) \in L_k \times L_k, I_k \vee J_k = 1\}|.$$

Let G be the bipartite graph on vertex sets L and R with $(G_{k-1}, G_{k+1}) \sim (I_k, J_k)$ iff there exists a point q with $G_{k-1} \vee q = J_k$, $G_{k+1} = I_k \vee q$. While (8.18) is not known in general, it can be shown that in G there is no matching from all of R into L.

Let us finally derive some sufficient conditions for (8.16). For $a \in L_k$ denote by $|a| = |\{p \in L_1 : p \leqslant a\}|$ the number of points below a and set $z(a) = |\{(p, b) : b \lessdot a, p \in L_1, b \vee p = a\}|$. Let d_k, D_k be the averages of these numbers, i.e.,

$$d_k = \frac{1}{W_k} \sum_{a \in L_k} |a| \quad k = 1, \dots, r,$$

$$D_k = \frac{1}{W_k} \sum_{a \in L_k} z(a) \quad k = 2, \dots, r. \tag{8.19}$$

8.5.8. Proposition. *Let L be a geometric lattice with the numbers d_k, D_k defined as in (8.19), $k \geqslant 2$. If $d_k \geqslant d_{k-1}$ and $D_{k+1} \geqslant D_k$ then $W_k^2 \geqslant W_{k-1} W_{k+1}$.*

Proof. We have

$$\sum_{a\in L_k} z(a) = \sum_{a\in L_k}\left(\sum_{\substack{b\\b<a}}(|a|-|b|)\right) = \sum_{b\in L_{k-1}}\left(\sum_{\substack{a\\a>b}}(|a|-|b|)\right) = \sum_{b\in L_{k-1}}(W_1 - |b|).$$

From $D_{k+1} \geqslant D_k$ we infer

$$\frac{1}{W_{k+1}}\sum_{a\in L_k}(W_1 - |a|) \geqslant \frac{1}{W_k}\sum_{b\in L_{k-1}}(W_1 - |b|)$$

and hence

$$\frac{W_k(W_1 - d_k)}{W_{k+1}} \geqslant \frac{W_{k-1}(W_1 - d_{k-1})}{W_k},$$

i.e., $W_k^2 \geqslant W_{k-1}W_{k+1}$ since $d_{k-1} \leqslant d_k < W_1$. $\square$

8.5.9. Example. Suppose the geometric lattice L has the property that for all $k = 1, \ldots, r$ any $a \in L_k$ contains the same number p_k of points [called a matroid design in Young, Murty, and Edmonds (1970)]. Then, trivially, $d_k = p_k$ and $d_k \geqslant d_{k-1} + 1$ for all k. By an inductive argument it is easy to see that any $a \in L_k$ also covers the same number c_k of elements, in fact,

$$c_k = \prod_{i=0}^{k-2}\frac{p_k - p_i}{p_{k-1} - p_i},$$

whence $z(a) = c_k(p_k - p_{k-1})$ depends only on k. From this it follows right away that $D_{k+1} \geqslant D_k + 1$. By examining the proof of Proposition 8.5.8 we may derive the even stronger inequality

$$\frac{W_k^2}{W_{k-1}W_{k+1}} \geqslant \frac{W_1 - k + 1}{W_1 - k} \quad (1 \leqslant k \leqslant r - 1). \tag{8.20}$$

That (8.20) holds for *all* geometric lattices was conjectured by Mason (1972) where he conjectured that even

$$\frac{W_k^2}{W_{k-1}W_{k+1}} \geqslant \frac{W_1 - k + 1}{W_1 - k}\frac{k + 1}{k} \quad (1 \leqslant k \leqslant r - 1) \tag{8.21}$$

may hold for all geometric lattices.

Inequality (8.21) has been verified for $k = 2$ in the case when any line contains at most four points (Seymour 1982).

8.6. Comments

To conclude, we make a few remarks on the original development of the various notions and results mentioned in this chapter. Related surveys that have appeared are Mason (1972) and Welsh (1976, Chapters 15, 16).

Section 8.2: The chromatic polynomial of a graph was first studied by Birkhoff and Lewis (1946) and Whitney (1932a and b) where it is proved that the corresponding Whitney numbers of the first kind are alternating in sign; see also Read (1968) and Tutte (1954). Rota gave the general definition of the characteristic polynomial of a lattice and extended the above results to arbitrary geometric lattices (Rota 1964). It was Rota also who proposed the unimodality conjectures (Rota 1970). The concept of an incidence algebra of a poset has its origin in the theory of arithmetic functions in number theory; for a thorough account see Doubilet, Rota, and Stanley (1972). The proof of Proposition 8.2.6 is taken from Comtet (1970).

Section 8.3: Möbius algebras were introduced by Solomon (1967) and studied further by several authors, particularly Greene (1971, 1973). The very elegant procedure outlined in (8.7) to Corollary 8.3.3 and in Proposition 8.3.8 is due to Dowling and Wilson (1975); Proposition 8.3.9 appears in Dowling (1977). Corollary 8.3.4 is known as Weisner's theorem, and Proposition 8.3.6 was proved in Rota (1964). For general results on the Möbius function see Rota (1964), Crapo (1966), Aigner (1979, Chapter 4), and Chapter 7.

Section 8.4: The inductive procedure used in the proof of Theorem 8.4.1 belongs to the field of arithmetical invariants in matroids, which originated with the work of Tutte in (1947). (See, e.g., Crapo 1969). In this setting, (8.9) says that ψ is a chromatic invariant. The inequalities in Corollary 8.4.3 were proved in Heron (1972), the characterization of the left-hand equality in Dowling and Wilson (1974). Refinements are due to Brylawski (1977) and Björner (1987) (see also White 1988).

Section 8.5: The hyperplane theorem $W_1 \leqslant W_{r-1}$ was proved by many authors, e.g. Basterfield and Kelly (1968), Greene (1970), Heron (1973), and Motzkin (1951), with the more precise matching results Proposition 8.5.3 (for $k = 1$) first appearing in Greene (1970) where, also, equality is characterized by the modularity of the lattice. The generalization Proposition 8.5.6 was given by Mason (1973). Our development Proposition 8.5.1 to Proposition 8.5.5 is due to Dowling and Wilson (1975) who also established the bounds in Proposition 8.5.5 and characterized the lattices for which equality holds (Dowling and Wilson 1974). Inequality 8.16 was proved for graphs by Stonesifer (1975); the more general results Proposition 8.5.7 and (8.21) are due to Aigner and Schoene (preprint) and Seymour (1982).

References

Aigner, M. (1979). *Combinatorial Theory.* Springer-Verlag, Berlin–Heidelberg–New York.
Aigner, M. and Schoene, J. On the logarithmic concavity of the Whitney numbers. Preprint.

Basterfield, J.G. and Kelly, L.M. (1968). A characterization of sets of n points which determine n hyperplanes. *Proc. Camb. Phil. Soc.* **64**, 585–8.

Birkhoff, G.D. and Lewis, D.C. (1946). Chromatic polynomials. *Trans. Amer. Math. Soc.* **60**, 355–451.

Björner, A. (1987). Inequalities between Whitney numbers.

Blackburn, J.A., Crapo, H.H. and Higgs, D.A. (1973). A catalogue of combinatorial geometries. *Math. Comp.* **27**, 155–66.

Brylawski, T.H. (1977a). Connected matroids with the smallest Whitney numbers. *Discrete Math.* **18**, 243–52.

Brylawski, T.H. (1977b). The broken-circuit complex. *Trans. Amer. Math. Soc.* **234**, 417–33.

Canfield, R. (1978). On a problem of Rota. *Bull Amer. Math. Soc.* **84**, 164.

Comtet, L. (1970). *Analyse Combinatoire.* Presses Univ. de France, Paris.

Crapo, H.H. (1966). The Möbius function of a lattice. *J. Comb. Theory* **1**, 126–31.

Crapo, H.H. (1969). The Tutte polynomial. *Aequationes Math.* **3**, 211–29.

Dilworth, R.P. and Greene, C. (1971). A counterexample to the generalization of Sperner's theorem. *J. Comb. Theory* **10**, 18–20.

Doubilet, P. Rota, G.-C. and Stanley, R.P. (1972). On the foundations of combinatorial theory. VI: The idea of generating function. *Proc. 6th Berkeley Symp. on Math. Stat. and Prob.*, vol. II: *Probability Theory*, 267–318.

Dowling, T.A. (1977). Complementing permutations in finite lattices. *J. Comb. Theory Ser. B* **23**, 223–6.

Dowling, T.A. and Wilson, R.M. (1974). The slimmest geometric lattices. *Trans. Amer. Math. Soc.* **196**, 203–15.

Dowling, T.A. and Wilson, R.M. (1975). Whitney number inequalities for geometric lattices. *Proc. Amer. Math. Soc.* **47**, 504–12.

Greene, C. (1970). A rank inequality for finite geometric lattices. *J. Comb. Theory* **9**, 357–64.

Greene, C. (1971). An inequality for the Möbius function of a geometric lattice, in *Proc. Conf. on Möbius algebra.* University of Waterloo.

Greene, C. (1973). On the Möbius algebra of a partially ordered set. *Advances Math.* **10**, 177–87.

Heron, A.P. (1972). Matroid polynomials, in *Combinatorics* (Institute of Math. and Appl., D.J.A. Welsh and D.R. Woodall, eds.), pp. 164–203.

Heron, A.P. (1973). A property of the hyperplanes of a matroid and an extension of Dilworth's theorem. *J. Math. Anal. Appl.* **42**, 119–32.

Mason, J.H. (1972). Matroids: Unimodal conjectures and Motzkin's theorem, in *Combinatorics* (Institute of Math. and Appl., D.J.A. Welsh and D.R. Woodall, eds.), pp. 207–21.

Mason, J.H. (1973). Maximal families of pairwise disjoint maximal proper chains in a geometric lattice. *J. London Math. Soc.* **6**, 539–42.

Motzkin, T. (1951). The lines and planes connecting the points of a finite set. *Trans. Amer. Math. Soc.* **70**, 451–64.

Read, R.C. (1968). An introduction to chromatic polynomials. *J. Comb. Theory* **4**, 52–71.

Rota, G.-C. (1964). On the foundations of combinatorial theory I: Theory of Möbius functions. *Z. Wahrscheinlichkeitsrechnung u. verw. Gebiete* **2**, 340–68.

Rota, G.-C. (1970). Combinatorial theory, old and new. *Proc. Int. Congress Math.* (*Nice*) **3**, 229–33.

Seymour, P.D. (1982). On the points–lines–planes conjecture. *J. Comb. Theory Ser. B* **33**, 17–26.

Solomon, L. (1967). The Burnside algebra of a finite group. *J. Comb. Theory* **2**, 603–15.

Stonesifer, J.R. (1975). Logarithmic concavity for edge lattices of graphs. *J. Comb. Theory Ser. A* **18**, 36–46.

Tutte, W.T. (1947). A ring in graph theory. *Proc. Camb. Phil. Soc.* **43**, 26–40.

Tutte, W.T. (1954). A contribution to the theory of chromatic polynomials. *Can. J. Math.* **6**, 80–91.

Welsh, D.J.A. (1976). *Matroid Theory.* Academic Press, London–New York–San Francisco.

White, N., ed. (1988). *Combinatorial Geometries: Advanced Theory*, Cambridge University Press, to appear.

Whitney, H. (1932a). A logical expansion in mathematics. *Bull. Amer. Math. Soc.* **38**, 572–9.

Whitney, H. (1932b). The coloring of graphs. *Annals Math.* **33**, 688–718.

Wille, R. (1971). On incidence geometries of grade *n*. *Atti del Convegno di Geom. Comb. e sue Appl*, pp. 421–6. University of Perugia.

Young, P., Murty, U.S.R. and Edmonds, J. (1970). Equicardinal matroids and matroid designs. *Proc. 2nd Chapel Hill Conf. on Comb. Math. and its Appl.*, University of North Carolina, 498–542.

9

Matroids in Combinatorial Optimization

ULRICH FAIGLE

Matroids enter combinatorial optimization problems at various levels. Whitney's (1935) motivation to introduce matroids as combinatorial objects in their own right stemmed from his interest in approaching the Four-Color Problem algebraically and combining the combinatorial and algebraic–geometric aspects of graphs into the notion of a matroid.

Graphs furnish the most important models for combinatorial optimization problems. Thus it is natural to ask to what extent graph properties actually are properties of the underlying matroid and to study more general classes of matroids that enjoy, for example, the 'max-flow-min-cut' property of network flows (cf. Seymour 1977). This approach leads to fundamental structural questions about matroids *per se* which, nevertheless, have many practical implications. One of the foremost results in this area is Seymour's (1980) decomposition theory for regular matroids exhibiting regular matroids as being essentially built up by graphic and cographic matroids. As a consequence, efficient procedures can be developed to test whether a matrix is totally unimodular or whether certain linear programs actually are (better tractable) network problems (see, e.g., Welsh 1982 and Bixby 1982 for an introduction into this aspect of matroid theory).

Matroids also compose the combinatorial structure of linear programming (Minty 1966, Rockafellar 1969). Indeed, pivoting in linear programming may be carried out purely 'combinatorially' (Bland 1977).

A third aspect brings in matroids not only as a combinatorial abstraction of optimization problems but as an essential tool in their combinatorial analysis. It is this aspect that we want to focus on here.

Many combinatorial optimization problems can be modeled as optimization problems over an independence system on a ground set E, which we will always assume to be finite (see Section 9.1 for details). Already Boruvka (1926) discovered that the greedy heuristic affords an optimal strategy if the independence system is the collection of independent sets of a matroid. The optimization problem is also tractable if the independence system in question is the intersection of two matroid independence systems. The classical example of such a problem is the matching problem on bipartite graphs. In fact, the matroid intersection algorithm we describe in Section 9.2 can be understood as a generalization of König's (1936) augmenting path technique to solve the bipartite matching problem. Here is where matroids enter essentially. Intersections of matroids generally do not result in matroids. The dual construction, matroid union, preserves the matroid property – but may lead out of a particular class of matroids. For example, the union of two graphic matroids need not be graphic itself. In other words, the framework of graphs does not capture this construction.

In Section 9.3 we introduce integral matroids. They are the collections of integral vectors of integral polyhedral matroids in the sense of Edmonds (1970). Another viewpoint allows the interpretation of integral matroids not just as integral points in certain convex polyhedra but as matroids on multisets. The framework of integral matroids is an appropriate means for many combinatorial optimization problems whose constraints are presented by integer-valued submodular set functions. Yet, the Dilworth completion, which generalizes Dilworth's construction for embedding arbitrary lattices into geometric lattices (cf. Crawley and Dilworth 1973, Chapter 14), reduces the theory of integral matroids to 'classical' matroid theory.

Submodular functions and supermodular functions determine fundamental discrete structures ranging from information theory (Fujishige 1978a) and game theory (cf. the survey of Rosenmüller 1983) and engineering (cf. Iri 1983) to cluster analysis (cf. Barthélemy, Leclerc, and Monjardet 1984).

They are the discrete analogs of convex and concave functions in non-linear optimization (see Lovász 1983, Fujishige 1984b). Concentrating on systems determined by integral sub- and supermodular functions, we will outline in Section 9.4 how their theory can be developed within the framework of integral matroids and hence of matroids.

The final Section 9.5 discusses the network flow model with submodular restrictions of Edmonds and Giles (1977). It comprises the network flow model of Ford and Fulkerson (1962) as a special case and we will spend time to take a close look at this model from a matroid point of view. We end with an efficient combinatorial algorithm to minimize a submodular function over the power set of the ground set E.

Our purpose here is not to provide the reader with a comprehensive introduction into the theory of combinatorial optimization. We want to exhibit the particular role matroids play within an analysis of combinatorial optimization problems. For a more detailed introduction into the general theory of combinatorial optimization and its applications, we refer the reader to, e.g., the textbooks of Lawler (1976) or Papadimitriou and Steiglitz (1982).

9.1. The Greedy Algorithm and Matroid Polyhedra

Let us consider a very general optimization problem in a combinatorial setting. There we are given a real-valued *weight* function $c: E \to \mathbb{R}$ on the ground set E and a (non-empty) family $\mathscr{F} \subseteq 2^E$ of subsets of E. For every subset $A \subseteq E$, we have the induced weight

$$c(A) = \sum_{a \in E} c(a), \qquad (9.1)$$

where, as usual, $c(\varnothing) = 0$ is understood. The optimization problem can now be stated as

$$\max c(F) \quad \text{subject to} \quad F \in \mathscr{F}. \qquad (9.2)$$

Note that we could equally well formulate the standard problem as

$$\min c(F) \quad \text{subject to} \quad F \in \mathscr{F}.$$

Indeed, maximizing c is equivalent to minimizing the negative weighting $(-c)$. The following three examples represent typical problems of this kind. We state them in the language of graph theory and denote by $\Gamma = \Gamma(V, E)$ a graph with vertex set V and edge set E. So $c: E \to \mathbb{R}$ is a weighting on the edges of Γ.

9.1.1. Minimum Spanning Tree. *$\mathscr{F}$ consists of the edge sets of the spanning trees of Γ. Determine one with the least weight.*

9.1.2. Traveling Salesman. *$\mathscr{F}$ consists of the edge sets of all closed paths of Γ which meet every vertex. Find a 'shortest' one.*

9.1.3. Bipartite Matching. *$\mathscr{F}$ consists of those edge sets of the bipartite graph Γ which only contain pairwise non-incident edges. Construct one with best possible weight.*

A problem of type (9.1.1), for instance, occurs when a communications network connecting all vertices has to be established and the weight $c(e)$ reflects the cost of a direct link between the two terminal vertices of the edge e. Problem (9.1.2) seems to be similar: every vertex of Γ has to be visited so that the total length of the tour is minimal, where $c(e)$ measures the distance between the two end vertices of the edge e. We have already encountered the third problem in Chapter 4 for the weight function $c \equiv 1$. We allow now for the possibility that different pairings of vertices in Γ may have different values.

Problem (9.1.2) is a representative of the class of so-called *NP-complete* problems (cf. Garey and Johnson 1979), which appear hard to solve efficiently. In spite of its similarity, however, 9.1.1 is 'easy' and also 9.1.3 is quite tractable as we will see. Without going too much into details, let us be a little more specific.

All three problems above can be solved with the following straightforward method: list all members of $\mathscr{F}$ together with their weights and select an optimal one. Since the ground set E is finite, this procedure will certainly terminate, after a finite amount of time, with the correct result. But is this method practical? In most cases, $\mathscr{F}$ will be prohibitively large. The complete graph K_n, for example, has n^{n-2} spanning trees (Cayley's theorem)! Thus procedures are called for which substantially reduce the amount of work. Here the amount of work is measured by the number of steps the execution of the procedure requires when implemented on an ideal computer. A (correct) procedure is said to be a *good* algorithm for a class of problems if this number of steps is bounded by a polynomial in the size of the problems (J. Edmonds). In this sense, no good algorithm is known for the Traveling Salesman Problem.

Let us return to the general optimization problem (9.2) and let us assume from now on that $\mathscr{F}$ is an *independence system*, i.e., for all $A, B \subseteq E$

$$B \in \mathscr{F} \quad \text{and} \quad A \subseteq B \quad \text{implies} \quad A \in \mathscr{F}.$$

Then we can try the following simple heuristic, which builds up a member of $\mathscr{F}$ 'greedily' from the empty set by adjoining in each step the best element currently available to the set already constructed.

9.1.4. Greedy Algorithm.
(i) *Order the elements of E so that $c(e_1) \geqslant c(e_2) \geqslant \cdots \geqslant c(e_n)$*;
(ii) $B \leftarrow \varnothing$;
(iii) $i \leftarrow 1$;
(iv) IF $B \cup \{e_i\} \in \mathscr{F}$ THEN $B \leftarrow B \cup \{e_i\}$;
(v) $i \leftarrow i + 1$;
(vi) IF $i \leqslant n$ THEN GOTO   (iv);
(vii) STOP.

It is easy to see that the greedy algorithm need not produce a solution to (9.1) even when the weight function $c: E \to \mathbb{R}_+$ is non-negative. What are the independence systems for which the greedy algorithm is optimal with respect to every non-negative weight function? It is interesting to note that the answer was given by the electrical engineer O. Boruvka (1926) before the birth proper of matroid theory. He essentially proved the following fundamental result.

9.1.5. Theorem. *The non-empty independence system $\mathscr{F} \subseteq 2^E$ of subsets of E is*

the collection of independent sets of some matroid on E if and only if, for every non-negative weight function $c: E \to \mathbb{R}^+$, the greedy algorithm 9.1.4 solves the optimization problem (9.1).

Proof. We show the necessity of the matroid property by verifying the basis exchange axiom for the collection $\mathcal{B} = \mathcal{B}(\mathcal{F})$ of maximal members of $\mathcal{F}$. Sufficiency will follow from the discussion in the remainder of this section.

For $A, B \in \mathcal{B}$ and $a \in A \backslash B$, we define a weighting $c: E \to \mathbb{R}_+$ by

$$c(e) = \begin{cases} 1 & \text{if } e \in (A \backslash \{a\}) \cup B \\ 0 & \text{otherwise.} \end{cases}$$

Ordering the elements of E so that $A \backslash \{a\}$ is in accordance with the greedy algorithm, we see that 9.1.4 can only be successful if there exists an element $b \in B \backslash A$ with $(A \backslash \{a\}) \cup \{b\} \in \mathcal{B}$. $\qquad\Box$

The non-negativity requirement for the weight function in Theorem 9.1.5, of course, is no real restriction. The greedy algorithm can always be adjusted in the obvious manner (see Exercise 9.1).

Boruvka's theorem has been rediscovered many times. For graphic matroids, the greedy algorithm yields Kruskal's (1956) solution of the minimum spanning tree problem 9.1.1. Rado (1957) gives a general matroid formulation. The full power of the greedy algorithm was realized by Edmonds (1971).

The greedy algorithm can be looked at as a purely combinatorial construction (cf. Exercise 9.2). Edmond's (1970) idea to set it in the framework of linear programming, however, helps to gain further insight into the structure of combinatorial optimization problems. Let us recall a few basic facts first (for more details see, e.g., Chvátal 1983).

With respect to the $(m \times n)$ matrix A and vectors $b \in \mathbb{R}^m$, $c \in \mathbb{R}^n$ we state the (*primal*) *linear program* (LP):

$$\max c \cdot x \quad \text{subject to } Ax \leqslant b \quad \text{and} \quad x \geqslant 0. \tag{9.3}$$

With the LP (9.3) we associate its *dual*:

$$\min b \cdot y \quad \text{subject to } yA \geqslant c \quad \text{and} \quad y \geqslant 0. \tag{9.4}$$

[Note that (9.4) may also be expressed in the form (9.3) and hence also is a LP].

The next lemma is fundamental because it gives an optimality criterion for linear programs.

9.1.6. Lemma. *Let the vector $x \in \mathbb{R}^n$ and the vector $y \in \mathbb{R}^m$ satisfy the restrictions of (9.3) and (9.4) respectively. Then*

$$c \cdot x \leqslant b \cdot y.$$

Hence, if $c \cdot x = b \cdot y$, x *and* y *must be optimal solutions to the respective LP's.*

Proof. Exercise 9.3. □

In our application to optimization problems over matroids, we consider $\mathbb{R}^E$ rather than just $\mathbb{R}^n$, that is, the space $\mathbb{R}^n$ with components indexed by the distinct elements of E. Equivalently, we could see $\mathbb{R}^E$ as the collection of all real-valued functions on E. This collection contains in particular the *characteristic* or (0, 1)-*incidence vectors* of the subsets of E, i.e., for every $A \subseteq E$ the vector χ_A, where

$$\chi_A(e) = \begin{cases} 1 & \text{if} \quad e \in A, \\ 0 & \text{otherwise.} \end{cases}$$

Furthermore, for $x \in \mathbb{R}^E$ and $A \subseteq E$, the notation $x(A)$ refers to the sum of all components of x with index in A as in (9.1).

So let M be a matroid on E with rank function r and collection $\mathscr{F}$ of independent sets. If $x = \chi_S$ for some $S \subseteq E$, we have

$$S \in \mathscr{F} \quad \text{if and only if} \quad x(A) \leqslant r(A) \quad \text{for all} \quad A \subseteq E.$$

Hence the following optimization problem generalizes (9.2):

$$\max c \cdot x \quad \text{subject to} \quad x \in \mathbb{P}(r), \tag{9.5}$$

where

$$\mathbb{P}(r) = \{ x \in \mathbb{R}^E : x \geqslant 0, \quad x(A) \leqslant r(A) \quad \text{for all} \quad A \subseteq E \}.$$

$\mathbb{P}(r)$ is called the *matroid polyhedron* (or *polymatroid* for short) associated with the matroid rank function r.

The reader should notice that (9.5) in fact is a LP of the form (9.3), where the rows of the matrix of restrictions are exactly the (0, 1)-incidence vectors of the subsets of E, whose ranks are the corresponding components of the vector b. We can now close the gap in the proof of Theorem 9.1.5.

9.1.7. Proposition. *Let $\mathscr{F}$ be the collection of independent sets of the matroid M with rank function $r : 2^E \to \mathbb{N}$ and let $c : E \to \mathbb{R}_+$ be non-negative. Then if $x^* \in \mathbb{R}^E$ is the characteristic vector of a greedy solution B^* of (9.2), x^* is an optimal solution of (9.5). Hence, also B^* is optimal.*

Proof. Because $x^* \in \mathbb{P}(r)$, it suffices, by Lemma 9.1.6, to exhibit an appropriate solution of the dual LP:

$$\min \sum_{A \subseteq E} r(A) \cdot y_A \quad \text{subject to} \quad y_A \geqslant 0 \quad \text{and, for all} \quad e \in E,$$

$$\sum \{ y_A : e \in A \} \geqslant c(e). \tag{9.6}$$

Assume the greedy algorithm 9.1.4 uses the ordering $c(e_1) \geqslant c(e_2) \geqslant \cdots \geqslant c(e_n)$.

to generate B^* and define the vector y^* indexed by subsets

$$y_A^* = \begin{cases} c(e_i) - c(e_{i-1}) & \text{if} \quad A = A_i \quad \text{for some } i = 1, \dots, n, \\ 0 & \text{otherwise,} \end{cases}$$

where we set $c(e_0) = 0$ and, for $i = 1, \dots, n$,

$$A_i = \{e_1, e_2, \dots, e_i\}.$$

It is not difficult to see that y^* satisfies the restrictions of (9.6) and that

$$c(B^*) = \sum_{A \subseteq E} r(A) \cdot y_A^*. \qquad \square$$

The set family $\mathscr{F}$ in the bipartite matching problem 9.1.1 generally is not a matroid independence system. Hence the procedure 9.1.4 need not generate an optimal solution [it is interesting, however, that already Monge (1781) solves certain assignment problems by the greedy algorithm (see also Derigs, Goecke and Schrader 1984)]. So the question arises: how good is the greedy heuristic 9.1.4 for general independence systems? An answer is provided by the approach of Jenkyns (1976) and Korte and Hausmann (1978), which we briefly outline.

With respect to the general independence system $\mathscr{F} \subseteq 2^E$, we define for every $A \subseteq E$ its

$$\text{rank} \quad r(A) = \max \{|F| : F \subseteq A, \quad F \in \mathscr{F}\}$$

and

$$l - \text{rank} \quad \rho(A) = \min \{|F| : F \in \mathscr{F}, \quad F \subseteq A \quad \text{and}$$
$$F \cup \{a\} \notin \mathscr{F} \quad \text{for all} \quad a \in A \backslash F\}.$$

Then we obtain

9.1.8. Proposition. *Let $c : E \to \mathbb{R}_+$ be non-negative and let B^* be an optimal solution of the optimization problem (9.2) over the arbitrary independence system $\mathscr{F} \subseteq 2^E$. Furthermore, let B_G be a greedy solution for (9.2) obtained from 9.1.4. Then*

$$\min_{A \subseteq E} \frac{\rho(A)}{r(A)} \leqslant \frac{c(B_G)}{c(B^*)} \leqslant 1.$$

Proof. Assuming $c(e_1) \leqslant c(e_2) \leqslant \cdots \leqslant c(e_n)$, we define, as in the proof of Proposition 9.1.7, for $i = 1, \dots, n$,

$$A_i = \{e_1, e_2, \dots, e_i\}.$$

Hence, with $c(e_{n+1}) = 0$, we obtain

$$c(B_G) = \sum_{i=1}^{n} |B_G \cap A_i| (c(e_i) - c(e_{i+1}))$$

and

$$c(B^*) = \sum_{i=1}^{n} |B^* \cap A_i|(c(e_i) - c(e_{i+1})).$$

By the definition of 9.1.4, $B_G \cap A_i$ is a maximal member of $\mathscr{F}$ contained in A_i, i.e., $\rho(A_i) \leqslant |B_G \cap A_i|$. Hence $r(A_i) \geqslant |B^* \cap A_i|$ implies

$$|B_G \cap A_i| \geqslant |B^* \cap A_i| \cdot \frac{\rho(A_i)}{r(A_i)} \geqslant |B^* \cap A_i| \cdot \min_{A \subseteq E} \frac{\rho(A)}{r(A)}.$$

Thus

$$c(B_G) \geqslant c(B^*) \cdot \min_{A \subseteq E} \frac{\rho(A)}{r(A)}. \qquad \square$$

If $\mathscr{F}$ is a matroid independence system, the rank functions r and ρ coincide. Proposition 9.1.8, therefore, provides another proof for the sufficiency condition in Theorem 9.1.5. This is a special case of the following general lower bound for the quotients in Proposition 9.1.8.

9.1.9. Proposition. *Let $\mathscr{F} = \cap_{i=1}^{k} \mathscr{I}_i$, where for $i = 1, \ldots, k$, $\mathscr{I}_i$ is the system of independent sets of a matroid M_i on E with matroid rank function r_i. Then*

$$\frac{1}{k} \leqslant \min_{A \subseteq E} \frac{\rho(A)}{r(A)}.$$

Proof. We proceed by induction on the cardinality $|E|$ of the ground set and consider an arbitrary subset $A \subseteq E$.

Choosing sets $B, B^l \in \mathscr{F}$ so that $B, B^l \subseteq A$ and $r(A) = |B|$ and $\rho(A) = |B^l|$, we may assume $B^l \nsubseteq B$.

Let $b \in B^l \setminus B$ and denote by $\mathscr{I}_1', \ldots, \mathscr{I}_k', \mathscr{F}'$ the systems induced by the contractions $M_1/b, \ldots, M_k/b$. Thus

$$\rho'(A \setminus \{b\}) = |B^l \setminus \{b\}| = \rho(A) - 1,$$
$$r'(A \setminus \{b\}) \geqslant r(A) - k,$$

where the last inequality follows by augmenting the set $\{b\}$ with elements from B with respect to each of the k matroids $M_1, \ldots, M_k$.

By induction, we have $r'(A \setminus \{b\}) \leqslant k \cdot \rho'(A \setminus \{b\})$ and hence $r(A) \leqslant k \cdot \rho(A)$. $\square$

Let us illustrate Proposition 9.1.9 with the bipartite matching problem. If $\Gamma = \Gamma(S_1, S_2; E)$ is a bipartite graph with disjoint vertex sets S_1 and S_2 and edge set $E \subseteq S_1 \times S_2$, we may assume that we are also given matroids $M_1 = M(S_1)$ and $M_2 = M(S_2)$ on the two vertex sets (in 9.1.3, M_1 and M_2 are free matroids). $M_i (i = 1, 2)$ induces a matroid $\overline{M_i}$ on E, where $I \subseteq E$ is *independent* if I is incident with an independent set of M_i of cardinality $|I|$. Matchings then are those edge sets which are independent in both $\overline{M_1}$ and $\overline{M_2}$. So Proposition

9.1.9 applies with $k = 2$. In other words, the greedy strategy 9.1.4 is guaranteed to yield at least 50% of the optimum for the bipartite matching problem.

One can do better than using the greedy heuristic in order to tackle the optimization problem (9.2) over the intersection of two matroid independence systems efficiently. We describe an efficient algorithm for an exact solution in the next section. Not too surprisingly, the greedy algorithm is an integral part of it.

What about independence systems that are intersections of three or more matroids? No efficient solution algorithms for optimization problems over such systems are known. In fact, we can represent the following problem, which is known to be NP-complete (cf. Garey and Johnson 1979), in this form.

In the directed graph $G = (V, E)$ with set V of nodes and set E of arcs, we want to find out whether G possesses a directed *Hamiltonian path*, i.e., a directed path meeting each node exactly once. Letting $\mathscr{I}_1$ consist of the circuit-free edge sets of G (considered as undirected graph), $\mathscr{I}_2$ comprise all arc sets such that no two tails are incident with the same node, and, similarly, $\mathscr{I}_3$ be the collection of arc sets with no pair of incident heads, then our problem reduces to deciding whether the equation

$$\max\left\{|I| : I \in \mathscr{I}_1 \cap \mathscr{I}_2 \cap \mathscr{I}_3\right\} = |V| - 1$$

is valid.

9.2. Intersections and Unions of Matroids

We will now describe an algorithm to solve the optimization problem (9.2) over an independence system which can be expressed as the intersection of two matroid independence systems. The algorithm is good in the sense of Section 9.1 provided we have procedures at hand allowing to check efficiently whether a given subset of the ground set is independent with respect to the two matroids (cf. Exercise 9.6).

For the 'classical' bipartite matching problem an algorithmic solution of König (1936) employs the following fundamental idea: rather than simply adjoining element after element until the solution is obtained, one proceeds from an object which is optimal among all objects of cardinality k to an optimal object of cardinality $k + 1$ and so on, where the transformation is carried out according to a suitable 'augmenting path' with respect to the former object.

The general matroid problem was solved by Edmonds (1968, 1979) and also has attracted other researchers (cf. Lawler 1975, Iri and Tomizawa 1975). Our exposition here is based on Frank (1981). We concentrate on describing how to find an 'augmenting path' that allows us to transform an optimal k-object into an optimal $(k + 1)$-object.

So let $\mathscr{F} = \mathscr{I}_1 \cap \mathscr{I}_2$, where $M_1 = (E, \mathscr{I}_1)$ and $M_2 = (E, \mathscr{I}_2)$ are matroids with rank functions r_1 and r_2 and systems $\mathscr{I}_1$ and $\mathscr{I}_2$ of independent sets.

Given $c: E \to \mathbb{R}$, we consider the problem

$$\max c(F) \quad \text{subject to} \quad F \in \mathcal{F}. \tag{9.7}$$

With the notation

$$\mathcal{F}^k = \{F \in \mathcal{F} : |F| = k\},$$

we will actually solve the seemingly stronger problem

$$\max c(F) \quad \text{subject to} \quad F \in \mathcal{F}^k. \tag{9.8}$$

[Adding a suitable constant as to make $c: E \to \mathbb{R}$ non-negative, however, it is easy to see that (9.7) and (9.8) are, in fact, equivalent].

Introducing further notation, we write $C(I, x)$ for the (unique!) fundamental circuit contained in $I \cup x$ whenever I is an independent set of the matroid $M(E, \mathcal{I})$ and $x \in E$ is such that $I \cup x \notin \mathcal{I}$.

Calling the set $I \in \mathcal{I}^k$ *c-maximal* in $\mathcal{I}^k$ if $c(J) \leqslant c(I)$ holds for every $J \in \mathcal{I}^k$, we now observe

9.2.1. Lemma. *Assume $B \in \mathcal{I}^k$ is c-maximal in $\mathcal{I}^k$ and $x_1, x_2, \ldots, x_l \in E \backslash B$ and $y_1, y_2, \ldots, y_l \in B$ are elements so that*

$$B \cup x_i \notin \mathcal{I} \quad \text{and} \quad y_i \in C(B, x_i), \tag{9.9}$$

$$c(x_i) = c(y_i), \tag{9.10}$$

$$c(y_i) = c(y_j) \quad \text{and} \quad i < j \text{ implies } y_i \notin C(B, x_j). \tag{9.11}$$

Then $B' = (B \backslash \{y_1, \ldots, y_l\}) \cup \{x_1, \ldots, x_l\}$ is c-maximal in $\mathcal{I}^k$.

Proof. We must show $B' \in \mathcal{I}^k$. Since the case $l = 1$ is obvious, let us assume $l > 1$ and choose y_i as the element which lexicographically minimizes $(c(y_j), j)$.

Then $i \neq j$ implies $y_i \notin C(B, x_j)$ since, otherwise, $y_i \in C(B, x_j)$ implies $c(y_i) \geqslant c(x_j) = c(y_j)$, i.e., $c(y_i) = c(y_j)$ by the choice of y_i, and hence $i < j$ contradicts (9.11) and $i > j$ contradicts the choice of y_i.

We now claim that $\bar{B} = (B \backslash y_i) \cup x_i$ satisfies the hypothesis of the lemma with respect to $\{y_1, y_2, \ldots, y_l\} \backslash y_i$ and $\{x_1, \ldots, x_l\} \backslash x_i$, which finishes the proof by induction on l.

For ease of notation, assume $i = 1$ and suppose $\bar{B} \cup x_2 \in \mathcal{I}$. But then we arrive at the contradiction $B \cup x_2 \in \mathcal{I}$ since, in view of $y_1 \in C(B, x_1)$, B and $\bar{B}$ generate the *same* closed set in the matroid $M = (E, \mathcal{I})$. Furthermore, we conclude from $y_1 \notin C(B, x_2)$ that, in fact, $C(B, x_2) = C(\bar{B}, x_2)$. So (9.9) must hold. Similarly, also (9.11) is verified. $\qquad\square$

Returning to the intersection problem, we state, setting $\mathcal{I}_{12}^k = \mathcal{I}_1^k \cap \mathcal{I}_2^k$,

9.2.2. Lemma. *Assume that $c_i: E \to \mathbb{R}$ $(i = 1, 2)$ are functions such that*

$c = c_1 + c_2$ and that $I \in \mathscr{I}^k_{12}$ is c_i-maximal in $\mathscr{I}^k_i$. Then I is c-maximal in $\mathscr{I}^k_{12}$.

$\square$

The algorithm to solve the problem (9.7) constructs, for $k = 0, 1, 2, \ldots,$ c-optimal members of $\mathscr{I}^k_{12}(=\mathscr{F}^k)$. We show how to carry out the step $k \to k+1$.

Thus, let $I \in \mathscr{I}^k_{12}$ and $c_i : E \to \mathbb{R}$ $(i = 1, 2)$ be such that the hypothesis of Lemma 9.2.2 is satisfied. We then set, for $i = 1, 2,$

$$m_i = \max \{c_i(y) : y \notin I \quad and \quad I \cup y \in \mathscr{I}_i\},$$
$$X_i = \{x \in E \setminus I : I \cup x \in \mathscr{I}_i \quad and \quad c_i(x) = m_i\}.$$

An auxiliary directed graph $G = (E, A)$ is now defined with the set E as set of nodes and the set $A \subseteq E \times E$ of arcs, where for all $x, y \in E$,

$$(x, y) \in A \quad if \quad I \cup x \notin \mathscr{I}_1, y \in C_1(I, x), \quad and \quad c_1(x) = c_1(y);$$
$$(y, x) \in A \quad if \quad I \cup x \notin \mathscr{I}_2, y \in C_2(I, x), \quad and \quad c_2(x) = c_2(y).$$

There are two cases to consider. The first case will deal with an augmenting path yielding a c-maximal member of $\mathscr{I}^{k+1}_{12}$. In the second case, we may modify the current weightings $c_i : E \to \mathbb{R}$ and then repeat the whole procedure until eventually either the first case occurs and an augmentation is possible or $\mathscr{I}^{k+1}_{12}$ is seen to be empty.

Noting that the arcs of G either enter I or leave I, depending on whether they are defined with respect to the first or to the second matroid, let us try to find a directed path U from some node in X_2 to some node in X_1. This can be accomplished, for example, by adding a new 'source node' s and arcs (s, x_2), $x_2 \in X_2$, to G and then using an efficient shortest path procedure (cf., e.g., Lawler 1976) to find shortest paths from s to all reachable nodes of G.

Case (i). There exists a path U with node sequence $(x_0, y_1, x_1, y_2, \ldots, y_l, x_l)$ from $x_0 \in X_2$ to $x_l \in X_1$ and U is as short as possible.

9.2.3. Lemma. $I' = (I \setminus \{y_1, \ldots, y_l\}) \cup \{x_0, x_1, \ldots, x_l\}$ and $c'_i = c_i (i = 1, 2)$ satisfy the conditions of Lemma 9.2.2. Moreover, $c(I') - c(I) = m_1 + m_2$.

Proof. Since $B = I \cup x_0$ is obtained according to the greedy algorithm from the c'_2-optimal $I \in \mathscr{I}^k_2$, B must be c'_2-maximal in $\mathscr{I}^{k+1}_2$.

Clearly, $B \cup x_i \notin \mathscr{I}_2$, $y_i \in C_2(B, x_i)$, and $c(x_i) = c(y_i)$ for $i = 1, 2, \ldots, l$. Furthermore, $c(y_i) = c(y_j)$ and $i < j$ must imply $y_i \notin C_2(B, x_j)$ since, otherwise, U would admit a shortcut from y_i to x_j and, therefore, not be minimal. Hence we conclude from Lemma 9.2.1 that I' is c'_2-maximal in $\mathscr{I}^{k+1}_2$.

The same argument with respect to U traversed in the reversed order now shows that I' also is c'_1-maximal in $\mathscr{I}^{k+1}_1$. $\square$

Thus, in Case (i), U is an augmenting path of the desired kind for $I \in \mathscr{I}^k_{12}$.

Case (ii). There is no path connecting X_2 with X_1.

Let T consist of those nodes of G that can be reached *via* a directed path from X_2, and set

$$\delta_1 = \min\{c_1(y) - c_1(x): I \cup x \notin \mathscr{I}_1, x \in T, y \in C_1(I, x) \setminus T\},$$
$$\delta_2 = \min\{m_1 - c_1(x): I \cup x \in \mathscr{I}_1, x \in T \setminus I\},$$
$$\delta_3 = \min\{c_2(y) - c_2(x): I \cup x \notin \mathscr{I}_2, x \notin T, y \in C_2(I, x) \cap T\},$$
$$\delta_4 = \min\{m_2 - c_2(x): I \cup \in \mathscr{I}_2, x \notin T \cup I\},$$

where $\min(\varnothing)$ is understood to be ∞.

9.2.4. Lemma. $\delta = \min\{\delta_1, \delta_2, \delta_3, \delta_4\} > 0.$

Proof. Consider δ_1, for instance. $y \in C_1(I, x)$ implies $c_1(y) \geqslant c_1(x)$ since I was c_1-maximal in $\mathscr{I}_1^k$. $c_1(y) = c_1(x)$ would reveal (x, y) as an arc of G and hence $y \in T$ if $x \in T$. Thus, we must have $\delta_1 > 0$.

For δ_2, note that $m_1 = c_1(x)$ would, in particular, give $x \in X_1$. So $x \in T$ would mean that there is a path from X_2 to X_1, contradicting the assumption of Case (ii).

δ_3 and δ_4 can be dealt with similarly.   □

9.2.5. Lemma. $I' = I$ *and* $c_i': E \to \mathbb{R}$ *satisfy the conditions of Lemma 9.2.2, where*

$$c_1'(x) = \begin{cases} c_1(x) + \delta & \text{if } x \in T, \\ c_1(x) & \text{if } x \notin T, \end{cases}$$

and

$$c_2'(x) = c(x) - c_1'(x).$$

Proof. We show that I' is c_1'-maximal in $\mathscr{I}_1^k$. The analogous statement about c_2' again can be verified similarly.

Suppose I is not c_1'-maximal in $\mathscr{I}_1^k$. By the optimality of the greedy algorithm (9.1.4), there must exist elements $x \notin I$ and $y \in I$ so that $c_1'(x) > c_1'(y)$.

If $I \cup x \in \mathscr{I}_1$, then necessarily $c_1'(y) = c_1(y)$ and $c_1'(x) = c_1(x) + \delta$ since the c_1-maximality of I in $\mathscr{I}_1^k$ yields $c_1(x) \leqslant c_1(y)$. In particular, $x \in T$. But $x \in T$ implies $\delta \leqslant \delta_2 \leqslant m_1 - c_1(x)$. Hence $c_1(y) \geqslant m_1$ reveals the inequality $c_1'(x) \leqslant c_1'(y)$, a contradiction.

If $I \cup x \notin \mathscr{I}_1$, we may assume $y \in C_1(I, x)$ since the greedy algorithm produces a c_1'-maximal member of $\mathscr{I}_1^k$ in such a way that in each step the weight of the element chosen does not exceed the weights of the elements chosen previously. Then, as before, $c_1'(y) = c_1(y)$ and $c_1'(x) = c_1(x) + \delta$. Whence $\delta \leqslant \delta_1 \leqslant c_1(y) - c_1(x)$ and $c_1'(x) \leqslant c_1'(y)$, a contradiction.   □

We thus have proved the validity of the following algorithmic solution of the optimization problem (9.7) with respect to the weight function $c: E \to \mathbb{R}$, the matroids $M_1 = (E, \mathscr{I}_1)$ and $M_2 = (E, \mathscr{I}_2)$, and the independence system $\mathscr{F} = \mathscr{I}_1 \cap \mathscr{I}_2$.

9.2.6. Weighted Matroid Intersection Algorithm.

(0) $k \leftarrow 0; I \leftarrow \varnothing; c_1 \leftarrow 0; c_2 \leftarrow c;$

(1) *Construct the auxiliary directed graph G with respect to* $c_1, c_2, I;$

(2) IF *Case (ii) occurs* THEN GOTO (8);

(3) *Find a shortest path* $U = (x_0, y_1, \ldots, y_l, x_l)$ *from* X_2 *to* $X_1;$

(4) $I \leftarrow (I \backslash \{y_1, \ldots, y_l\}) \cup \{x_0, x_1, \ldots, x_l\};$

(5) *Output* $k, I;$

(6) $k \leftarrow k + 1;$

(7) GOTO (1);

(8) IF $\delta = \infty$ THEN GOTO (11);

(9) $c_1 \leftarrow c_1'$ *and* $c_2 \leftarrow c_2'$ *as in Lemma 9.2.5;*

(10) GOTO (1);

(11) STOP.

Algorithm (9.2.6) generates, for $k = 1, 2, \ldots$, a c-maximal member of $\mathscr{I}_{12}^k$ if one exists. Indeed, if $\delta = \infty$ is attained, none of the quantities $\delta_1, \delta_2, \delta_3, \delta_4$ is defined and, hence,

$$r_1(T) = |I \cap T| \quad \text{and} \quad r_2(E \backslash T) = |I \backslash T|.$$

Since, for all $I' \in \mathscr{I}_1 \cap \mathscr{I}_2$ and $S \subseteq E$,

$$|I'| \leqslant r_1(S) + r_2(E \backslash S)$$

generally holds (cf. also Corollary 9.2.9 below), I must have maximal cardinality in $\mathscr{F} = \mathscr{I}_{12}$. The finiteness of the algorithm can be seen as follows.

Whenever Case (i) occurs, the size of $|I|$ is increased by one, i.e., Case (i) occurs at most $|E|$ times. Assume now that Case (ii) occurs twice in a row. Then the new auxiliary graph G' will contain all the arcs of the old graph G that only involve nodes of T. Because $X_2' \supseteq X_1$, the property $\delta_1 > 0$ shows that the new T' will *strictly* include T. In other words, Case (ii) cannot occur more than $|E|$ times in a row.

Like the greedy algorithm for matroids, the matroid intersection algorithm has a natural setting in the context of linear programming. For this discussion we will retain the same notation as before and start with some observations about possible modifications of the algorithm.

If Case (ii) occurs, we may update the current weightings c_1 and c_2 with any number $\delta', 0 \leqslant \delta' \leqslant \delta$, without affecting the validity of Lemma 9.2.5. In particular, $\delta' = \min\{\delta, m_2\}$ is permissible as long as $m_2 \geqslant 0$.

Since the algorithm starts with $c_1 = 0$, we have $m_1 = 0$ at any stage of the algorithm regardless whether Case (i) or Case (ii) has occurred.

If we are just interested in a solution to the optimization problem (9.7) we can stop the algorithm as soon as $m_2 \leqslant 0$ or m_2 is not defined.

Assume now that $I^* \in \mathscr{I}_1 \cap \mathscr{I}_2$, $|I^*| = k$, is the optimal solution to (9.7) generated by the algorithm 9.2.6 after k augmentations. If, at this stage, m_2 is no longer defined, let c_1^* and c_2^* denote the current weightings. If $m_2 \geqslant 0$, carry out the updating of Case (ii) with $\delta' = \min\{\delta, m_2\}$ so that $m_2' = 0$ is achieved, and let c_1^* and c_2^* denote the weightings after this update.

9.2.7. Lemma. *Under the conditions above, $c_1^* \geqslant 0$. Moreover, for every $x \in E$,*

$$c_2^*(x) = \begin{cases} 0 & \text{if } x \in E \setminus I^* \quad \text{and} \quad I^* \cup x \in \mathscr{I}_2, \\ \geqslant 0 & \text{if } x \in I^*. \end{cases}$$

Proof. Since $c_1^* \geqslant 0$ is immediate from the definitions, we only verify the second statement.

If $I \in \mathscr{I}_1^{k-1} \cap \mathscr{I}_2^{k-1}$ denotes the predecessor of I^* during the algorithm, we must have $m_2 \geqslant 0$ at that stage (otherwise no augmentation would be carried out). Thus $c_2(x) < 0$ for some $x \in I^*$ would imply $x \notin I$ (and hence contradict $m_2 \geqslant 0$) or show that I was not c_2-maximal in $\mathscr{I}_2^{k-1}$, which is impossible.

Suppose $c_2^*(x) < 0$ for some $x \in I^*$. Then we must have had $c_2(x) < \delta' \leqslant m_2$ and hence $c_2(I^* \cup y) = c_2(I^*) + m_2 > c_2(I^*)$ for some $y \notin I^*$ with $I^* \cup y \in \mathscr{I}_2^{k-1}$, which violates the assumed c_2-maximality of I^* in $\mathscr{I}_2^k$. $\square$

Consider now the primal linear program [cf. also (9.5)]:

$$\max c \cdot x \text{ subject to } \begin{pmatrix} A \\ A \end{pmatrix} x \leqslant \begin{pmatrix} r_1 \\ r_2 \end{pmatrix} \quad \text{and} \quad x \geqslant 0, \qquad (9.12)$$

where A is the $(0, 1)$-matrix of the characteristic vectors associated with the subsets of E. (9.12) has the dual

$$\min \sum_{S \subseteq E} y_1(S) r_2(S) + \sum_{S \subseteq E} y_2(S) r_2(S) \qquad (9.13)$$

$$\text{subject to } (y_1, y_2) \begin{pmatrix} A \\ A \end{pmatrix} \geqslant c \quad \text{and} \quad y_1, y_2 \geqslant 0.$$

We want to show not only that (the characteristic vector of) I^* is an optimal solution to (9.12) but also how the weightings c_1^* and c_2^*, defined above, yield an optimal solution to (9.13). For simplicity, we assume here that neither $M_1 = (E, \mathscr{I}_1)$ nor $M_2 = (E, \mathscr{I}_2)$ has a loop.

Order the elements of $I^* = \{e_1, \ldots, e_k\} = \{f_1, \ldots, f_k\}$ according to $c_1^*(e_1) \geqslant \cdots \geqslant c_1^*(e_k) \geqslant 0$ and $c_2^*(f_1) \geqslant \cdots \geqslant c_2^*(f_k) \geqslant 0$ and set $c_1^*(e_{k+1}) = c_2^*(f_{k+1}) = 0$. Denoting by cl_1 and cl_2 the closure operators of M_1 and M_2, let furthermore

$$E_i = \mathrm{cl}_1\{e_1, \ldots, e_i\} \quad \text{and} \quad F_i = \mathrm{cl}_2\{f_1, \ldots, f_i\} \quad (i = 1, \ldots k).$$

The vectors $y_1^*, y_2^* \geqslant 0$ can then be defined *via*, for $S \subseteq E$,

$$y_1^*(S) = \begin{cases} c_1^*(e_i) - c_1^*(e_{i+1}) & \text{if } S = E_i, \quad 1 \leqslant i \leqslant k, \\ 0 & \text{otherwise,} \end{cases}$$

$$y_2^*(S) = \begin{cases} c_2^*(f_i) - c_2^*(f_{i+1}) & \text{if } S = F_i, \quad 1 \leqslant i \leqslant k, \\ 0 & \text{otherwise.} \end{cases}$$

To verify that y_1^* and y_2^* satisfy the restrictions of (9.13), we must check for all $x \in E$,

$$\sum \{y_1^*(S) + y_2^*(S) : x \in S, \quad S \subseteq E\} \geqslant c(x),$$

which, with the help of Lemma 9.2.7 is not very difficult to do and, therefore, left to the reader. Moreover, we compute

$$c(I^*) = \sum_{S \subseteq E} y_1^*(S) r_1(S) + \sum_{S \subseteq E} y_2^*(S) r_2(S) \tag{9.14}$$

and thus conclude the desired optimality from Lemma 9.1.6.

Of particular importance is the case where the objective function $c : E \to \mathbb{R}$ is *integral*, that is, takes on integer values only. Our discussion shows that in this case also c_1 and c_2 remain integral. In particular, the linear program (9.13) affords an integral solution. In the terminology introduced by Edmonds and Giles (1977), the linear program (9.12), therefore, is *totally dual integral*, i.e., admits an integral dual solution whenever the primal program has an integral objective function (and, of course, solutions exist at all).

Let us remark that a totally dual integral linear program of the form (9.3) necessarily also has an integral (primal) solution (no matter whether the weight vector $c \in \mathbb{R}^n$ is integral or not) if an optimal solution exists. This follows from Hoffman's (1974) theorem which says that a linear program admits integral optimal solutions if for every *integral* $c \in \mathbb{R}^n$, the optimal *value* $\max c \cdot x$ is an integer.

We combine the discussion of the matroid intersection algorithm into a result, due to Edmonds (1970).

9.2.8. Theorem. *Let* $\mathbb{P}(r_1)$ *and* $\mathbb{P}(r_2)$ *be two matroid polyhedra in* $\mathbb{R}^E$. *Then the linear program*

$$\max c \cdot x \quad \text{subject to} \quad x \in \mathbb{P}(r_1) \cap \mathbb{P}(r_2)$$

is totally dual integral. Moreover, the vertices of the polytope $\mathbb{P}(r_1) \cap \mathbb{P}(r_2)$ *are integral.* ☐

Totally dual integral systems often arise from combinatorial structures for which the min-max theorem of linear programming (Lemma 9.1.3) then turns into a combinatorial min-max result. For more general information we refer the reader to the survey articles of Edmonds and Giles (1984) and Schrijver (1983, 1984).

An important special case of Theorem 9.2.8 is the *matroid intersection theorem*, which we implicitly have already stated above:

9.2.9. Corollary. *Let $M_1 = (E, \mathscr{I}_1)$ and $M_2 = (E, \mathscr{I}_2)$ be two matroids with rank functions r_1 and r_2. Then*

$$max\{|I|: I \in \mathscr{I}_1 \cap \mathscr{I}_2\} = min\{r_1(S) + r_2(E \backslash S): S \subseteq E\}. \qquad (9.15)$$

Proof. The left-hand side of (9.15) is the LP (9.12) with $c \equiv 1$. The right-hand side follows from (9.13), using the fact that optimal solutions can be required to be integral. (9.14) furnishes the claimed equality.   □

The intersection of two matroid independence systems generally is not a matroid independence system. However, a dual construction, the *union* or *sum* of two matroids, always results in a matroid. In the same sense as matroid intersection may be understood as a statement about *covering* the ground set E minimally [cf., e.g., the right-hand side of (9.15), the special case of (9.13)], the union of matroids allows us to analyze *packing* problems with matroids. The latter, indeed, initiated this construction (cf. Nash-Williams 1964 and Edmonds 1965).

We define the *union* or *sum* of the matroids $M_1 = (E, \mathscr{I}_1)$ and $M_2 = (E, \mathscr{I}_2)$ to be the matroid $M = (E, \mathscr{I})$ [also denoted by $M = M_1 \vee M_2 = (E, \mathscr{I}_1 \vee \mathscr{I}_2)$] where

$$\mathscr{I} = \mathscr{I}_1 \vee \mathscr{I}_2 = \{I_1 \cup I_2 : I_1 \in \mathscr{I}_1, I_2 \in \mathscr{I}_2\}.$$

Note that we have not yet shown that $M_1 \vee M_2$ is a matroid. Our next fundamental lemma will also be of interest for the construction of the Dilworth completion in Section 9.3.

9.2.10. Lemma. *Let $\mathscr{F}$ be a collection of subsets of E such that $\varnothing \in \mathscr{F}$ and $A \cup B \in \mathscr{F}$ and $A \cap B \in \mathscr{F}$ whenever $A, B \in \mathscr{F}$. Furthermore, let $f: \mathscr{F} \to \mathbb{N}$ be a submodular function with $f(\varnothing) = 0$. Then the function r, given by*

$$r(S) = min\{f(X) + |S \backslash X| : X \in \mathscr{F}\}$$

for all $S \subseteq E$, is a matroid rank function.

Proof. r clearly has the unit increase property. For the submodularity of r, consider $S, T \subseteq E$ and choose $X, Y \in \mathscr{F}$ so that

$$r(S) = f(X) + |S \backslash X| \quad \text{and} \quad r(T) = f(Y) + |T \backslash Y|.$$

Then

$$|S \backslash X| + |T \backslash Y| \geqslant |(S \cup T) \backslash (X \cup Y)| + |(S \cap T) \backslash (X \cap Y)|$$

implies

$$\begin{aligned} r(S) + r(T) &\geqslant f(X \cup Y) + f(X \cap Y) + |S \backslash X| + |T|Y| \\ &\geqslant f(X \cup Y) + |(S \cup T) \backslash (X \cup Y)| \\ &\quad + f(X \cap Y) + |(S \cap T) \backslash (X \cap Y)| \\ &\geqslant r(S \cup T) + r(S \cap T). \end{aligned}$$
                                                                                    □

9.2.11. Theorem. *If M_1 and M_2 are matroids with rank functions r_1 and r_2 then $M_1 \vee M_2$ is a matroid with rank function*

$$r(S) = \min \{r_1(X) + r_2(X) + |S \setminus X| : X \subseteq E\}.$$

Proof. Denoting by $\mathscr{I}$ the collection of independent sets with respect to the matroid rank function r, i.e.,

$$\mathscr{I} = \{I \subseteq E : |I \cap S| \leqslant r(S) \text{ for all } S \subseteq E\},$$

it is straightforward to verify $\mathscr{I}_1 \vee \mathscr{I}_2 \subseteq \mathscr{I}$. We show $\mathscr{I}_1 \vee \mathscr{I}_2 \supseteq \mathscr{I}$.

Let $B \in \mathscr{I}$ be arbitrary. We must find $I_1 \in \mathscr{I}_1$ and $I_2 \in \mathscr{I}_2$ so that $B \subseteq I_1 \cup I_2$. To do this, we restrict M_1 and M_2 if necessary so that we can assume $B = E$, and consider the Whitney dual M_2^* of M_2 with rank function r_2^*. Then for every $S \subseteq E (= B)$,

$$\begin{aligned}
r_1(S) + r_2^*(E \setminus S) &= r_1(S) + r_2(S) + |E \setminus S| - r_2(E) \\
&\geqslant r(E) - r_2(E) \\
&= |E| - r_2(E) \\
&= r_2^*(E).
\end{aligned}$$

Hence, by (9.15), there exists a basis I_2^* of M_2 so that $I_2^* \in \mathscr{I}_1 \cap \mathscr{I}_2$. Thus $I_1 = I_2^*$ and $I_2 = E \setminus I_2^*$ yield a partition of B into sets $I_1 \in \mathscr{I}_1, I_2 \in \mathscr{I}_2$. $\qquad\square$

9.2.12. Corollary. *Let $M_1 = (E, \mathscr{I}_1), \dots, M_k = (E, \mathscr{I}_k)$ be k matroids with rank functions $r_1, \dots, r_k$. Then the matroid sum $M = M_1 \vee \cdots \vee M_k$ has the rank function r given by*

$$r(S) = \min \left\{ \sum_{i=1}^{k} r_i(X) + |S \setminus X| : X \subseteq E \right\}. \tag{9.16}$$

Proof. We proceed by induction on k and just show the case $k = 3$.

In view of Theorem 9.2.11, there are sets $X, Y \subseteq E$ so that

$$r(S) = r_1(Y) + r_2(Y) + |X \setminus Y| + r_3(x) + |S \setminus X|,$$

where we may assume $Y \subseteq X \subseteq S$. Thus

$$\begin{aligned}
r(S) &= r_1(Y) + r_2(Y) + r_3(X) + |S \setminus Y| \\
&\geqslant r_1(Y) + r_2(Y) + r_3(Y) + |S \setminus Y|.
\end{aligned}$$

On the other hand, given $T \subseteq E$, we choose disjoint sets $I_1 \in \mathscr{I}_1, I_2 \in \mathscr{I}_2, I_3 \in \mathscr{I}_2$ with $I_1 \cup I_2 \cup I_3 \subseteq S$ and $|I_1| + |I_2| + |I_3| = r(S)$. Then

$$\begin{aligned}
r(S) &= |I_1 \cap T| + |I_2 \cap T| + |I_3 \cap T| + |(I_1 \cup I_2 \cup I_3) \setminus T| \\
&\leqslant r_1(T) + r_2(T) + r_3(T) + |S \setminus T|.
\end{aligned}$$

So (9.16) must hold. $\qquad\square$

Immediate consequences are the *matroid partitioning theorems* of Edmonds (1965) and Nash-Williams (1964):

9.2.13. Corollary. *Let* $M = (E, \mathscr{I})$ *be a matroid with rank function* r. *Then* E *can be covered with* k *or less independent sets if and only if for every* $S \subseteq E$,

$$|S| \leqslant k \cdot r(S).$$

Moreover, M *has at least* k *pairwise disjoint bases if and only if for all* $S \subseteq E$,

$$k \cdot (r(E) - r(S)) \leqslant |E \backslash S| \qquad \qquad \square$$

In the special case of graphic matroids, we obtain the results of Nash-Williams (1961, 1964) and Tutte (1961). Let $G = (V, E)$ be an undirected graph. Then the minimum number of forests needed to cover E is equal to

$$\max \left[\frac{|<W>|}{|V| - 1} \right], \qquad (9.17)$$

where the maximum ranges over all subsets $W \subseteq V$ with $|W| \geqslant 2$ and $\langle W \rangle$ denotes the set of edges of the subgraph induced by W. If G is connected, the maximum number of pairwise disjoint spanning trees is equal to

$$\min \left[\frac{|E \backslash S|}{\kappa(V, S) - 1} \right], \qquad (9.18)$$

where the minimum ranges over all subsets $S \subseteq E$ so that the number $\kappa(V, S)$ of the graph (V, S) is at least two.

At the end of Section 9.1, we observed that the intersection problem may be computationally very hard if more than two matroids are involved.

Note that this is *not* the case for the matroid partitioning problem. The proof of Theorem 9.2.11 reduces the problem to partition a subset $B \subseteq E$ into disjoint sets I_1 and I_2 which are independent with respect to two prescribed matroids to a matroid intersection problem. To test whether the arbitrary subset $S \subseteq E$ can, for example, be partitioned into sets $I_i \in \mathscr{I}_1, i = 1, 2, 3$, where $M_i = (E, \mathscr{I}_i)$ are prescribed matroids, we first try to decompose S with respect to $M_1 \vee M_2$ and M_3, say, to obtain $S = I_{12} \cup I_3$ and then decompose I_{12} with respect to M_1 and M_2, etc.

As a consequence, we observe that there are good algorithms to solve the optimization problems whose optimal values are described by (9.17) and (9.18).

The matroid intersection problem admits a generalization to the *matroid matching problem* (Lovász 1980) or *matroid parity problem* (Lawler 1976). Here we start with a submodular function f on a set E which is 'almost' a matroid rank function. More precisely, we assume $f: 2^E \to \mathbb{N}$ to satisfy for all $A, B \subseteq E$ and $b \in E$,

$$f(\varnothing) = 0,$$

$$f(A) \leqslant f(A \cup b) \leqslant f(A) + 2,$$
$$f(A \cup B) + f(A \cup B) \leqslant f(A) + f(B).$$

It will follow from Theorem 9.3.1 in the next section that all set functions f which satisfy the above conditions are constructible in the following fashion.

Let M be a matroid with rank function r on some set E', E a subset of the collections of points and lines of M, and define $f: 2^E \to \mathbb{N}$ via, for all $A \subseteq E$,

$$f(A) = r(\bigcup A).$$

Under the stated conditions, the *matroid matching problem* is

$$\max |X| \quad \text{subject to} \quad f(X) = 2 \cdot |X|.$$

As an illustration, let us look at two examples.

9.2.14. Matroid Intersection. *If* $M_1 = (E, I_1)$ *and* $M_2 = (E, I_2)$ *are two matroids with rank functions* r_1 *and* r_2, *then we have for every* $X \subseteq E$,

$$X \in I_1 \cap I_2 \quad \text{iff} \quad 2 \cdot |X| = r_1(X) + r_2(X).$$

In other words, the intersection problem reduces to the matching problem for $f = r_1 + r_2$.

9.2.15. Graph Matching. *Consider a (not necessarily bipartite) graph* $G = (V, E)$ *and let for every* $A \subseteq E$,

$$f(A) = |V(A)|,$$

where $V(A)$ *denotes the collections of vertices of the subset A of edges of G. Thus the edge sets* $X \subseteq E$ *with the property*

$$f(X) = 2 \cdot |X|$$

cannot contain any pair of adjacent edges in G, i.e., are matchings in a general sense.

For the *matroid parity problem*, we are given a matroid $M = (E, I)$ with rank function r on the set E, $|E| = 2k$, where E is partitioned into k pairs A_i, $|A_i| = 2$, of elements of E:

$$E = A_1 \cup A_2 \cup \cdots \cup A_k.$$

Among all independent sets of M which can be expressed as unions of suitable A_i's we are to find one with the largest possible cardinality. We leave it to the reader to formulate 9.2.14 and 9.2.15 as matroid parity problems.

Like matroid intersection graph matching also affords an efficient combinatorial optimization algorithm (Edmonds 1965, see also, e.g., Lawler 1976). Can one hope for an efficient algorithm that solves the general matroid matching problem? The answer is negative. Any solution algorithm for the matroid matching problem will generally require an exponential number of steps with respect to $|E|$ (Lovász 1980, Jensen and Korte 1982). We will outline the argument in the setting of the matroid parity problem.

Assume $k = 2m$ is an even number and define two matroids M_1 and M_2 on E, $|E| = 2k$, as follows:

(1) The bases of M_1 consist exactly of those k-element subsets of E that cannot be expressed as unions of m of the pairs A_i.

(2) Fix m distinct pairs A_i and let B denote their union. The bases of M_2 then are the set B and all bases of M_1.

It is straightforward to check that M_1 and M_2 are indeed matroids and that the parity problem yields $2(m-1)$ as an optimal solution for M_1 but $2m$ as an optimal solution for M_2. Hence every correct combinatorial matroid parity algorithm in particular must be able to distinguish M_1 from a matroid of type M_2.

We now have to specify how matroids are presented in our computational model. Suppose this is done via a *basis oracle*, i.e., a subroutine which computes for any subset offered in the course of the algorithm whether or not it is a basis of the matroid under investigation.

Then $\binom{2m}{m}$ calls to the oracle are needed to make sure that the collection of bases is that of M_1 and not that of a matroid of the form M_2, which implies an exponential lower bound.

Although the matroid matching problem appears to be generally intractable, there are subclasses of problems that admit efficient solutions. We have seen this for matroid intersection in this section. Lovász (1981) is able to derive a polynomial algorithm for the cardinality matching problem in the case where the underlying matroid is linear *and* a linear representation is explicitly available. Tong, Lawler and Vazirani (1984) observe that the weighted parity problem for gammoids can be reduced to the weighted matching problem on graphs and, therefore, is polynomially solvable.

9.3. Integral Matroids

Many combinatorial optimization problems call for an extension of the matroidal model that we used in the last section. Consider, for example, the bipartite graph $G = (S, T; E)$ with disjoint sets S and T of nodes and set E of edges. Furthermore, suppose we are given vectors $a \in \mathbb{N}^S$ and $d \in \mathbb{N}^T$. Thinking of S and T as sets of 'supply' and 'demand' nodes, where $a(s)$ denotes the amount of the commodity in question available at $s \in S$ and, similarly, $d(t)$ is the demand at $t \in T$, we may ask whether the demand can be satisfied with 'flows' along edges of E and, if so, how it can be done. Moreover, a capacity vector

$c \in \mathbb{N}^E$ may have to be observed limiting the flow capacity of the edge $e \in E$ to $c(e)$.

This problem is closely related to the matching problem in bipartite graphs. Indeed, it reduces to the matching problem if $a \equiv 1, d \equiv 1$, and $c \equiv 1$, which is a consequence of the results of Section 9.2: linear optimization over the intersection of two matroid polyhedra always admits an integral optimal solution. As in the matching problem, we associate with each $U \subseteq T$ the function value

$f(U) = $ maximum total amount of flow possible from S into U.

Thus we obtain the condition

$$f(U) \geqslant d(U) \quad \text{for all} \quad U \subseteq T$$

as a necessary generalization of the condition in Hall's marriage theorem. It is not difficult to check that $f(U)$ is submodular on the subsets of T, but need not be a matroid rank function.

In this section, we will study systems that are determined by submodular functions. Edmonds (1970) introduced such systems as *polymatroids* in $\mathbb{R}^E$, i.e., 'matroid' polyhedra, where the defining submodular function need not be unit-increasing, while Helgason (1974) investigated *hypermatroids* as generalized matroids on E with a submodular rank function lacking the unit-increase property. As we will see, the power of polymatroids and hypermatroids lies in the equivalence of these concepts.

We say that $f : 2^E \to \mathbb{N}$ is a *ground set rank function* if for all $A, B \subseteq E$,

$$f(\varnothing) = 0,$$
$$A \subseteq B \quad \text{implies} \quad f(A) \leqslant f(B),$$
$$f(A \cup B) + f(A \cap B) \leqslant f(A) + f(B).$$

(E, f) then is a *hypermatroid* on E. With the hypermatroid (E, f) we associate the *integral polymatroid* (or *integral matroid* for short)

$$Q(f) = \{x \in \mathbb{N}^E : x(S) \leqslant f(S) \quad \text{for all} \quad S \subseteq E\}, \quad (9.19)$$

where $x(S) = \sum_{s \in S} x(s)$ is the sum of the components of the vector x with index in S.

Note that each $x \in \mathbb{N}^E$ can be viewed as a *multiset* on E, where the component $x(e)$ indicates the *multiplicity* of the element $e \in E$ occurring in x.

A third aspect is order theoretic. Let $c \in \mathbb{N}^E$ be a *bounding vector* for $Q(f)$, that is, for all $x \in Q(f)$

$$x \leqslant c, \text{ i.e.,} \quad x(e) \leqslant c(e) \quad \text{for all} \quad e \in E$$

[for instance, every vector $c' \in \mathbb{N}^E$ with $c'(e) \geqslant f(e)$ for all $e \in E$ is bounding since f is a ground set rank function]. There is a natural correspondence between the e-th component of c and the set

$$C(e) = \{1(e) < 2(e) < \cdots < |c(e)|(e)\}$$

with $|c(e)|$ elements $[C(e) = \varnothing$ if $c(e) = 0]$. The disjoint union of the chains $C(e), e \in E$,

$$P(c) = \bigcup_{e \in E} C(e)$$

is ordered by the induced ordering. Thus every vector $x \leqslant c$ may be interpreted as an order ideal of $P(c)$ (and conversely) in the obvious way. [Recall that $I \subseteq P(c)$ is an *order ideal* if $a \in I$ and $b \in P(c)$ implies $b \in I$ whenever $b \leqslant a$.] Unions and intersections of order ideals are order ideals and reflect the vector operations for $x, y \in \mathbb{N}^E$,

$$x \vee y = (\ldots, \max\{x(e), y(e)\}, \ldots),$$

$$x \wedge y = (\ldots, \min\{x(e), y(e)\}, \ldots).$$

A special role is played by the collection $\mathscr{F}(c)$ of all order ideals $C(A), A \subseteq E$, of $P(c)$ of the form

$$C(A) = \bigcup_{e \in A} C(e).$$

The ground set rank function f naturally extends to $\mathscr{F}(c)$ *via*

$$f(C(A)) = f(A) \quad \text{for all} \quad A \subseteq E.$$

Hence Lemma 9.2.10 yields the *Dilworth embedding* $M(f; c)$ of the integral matroid $Q(f)$:

9.3.1. Theorem. *With respect to the integral matroid $Q(f) \subseteq \mathbb{N}^E$ and the bounding vector $c \in \mathbb{N}^E$ for $Q(f)$,*

$$r(S) = \min\{f(C(A)) + |S \setminus C(A)| : A \subseteq E\},$$

$S \subseteq P(c)$, is the rank function of a matroid $M = M(f; c)$ on $P(c)$. Moreover, the ideals of $P(c)$ corresponding to vectors of $Q(f)$ are exactly those ideals of $P(c)$ that are independent in $M(f; c)$.

Proof. In view of Lemma 9.2.10 and the definition (9.19) of $Q(f)$, it remains to show that for every independent ideal $X \subseteq P(c)$ with corresponding vector $x \in \mathbb{N}^E, x \leqslant c$, we have $x \in Q(f)$.

From $r(X) = |X|$, however, we immediately conclude

$$x(A) = |X \cap C(A)| = r(X) - |X \setminus C(A)| \leqslant f(A). \qquad \square$$

For the members of $\mathscr{F}(c)$, the rank in $M(f; c)$ is just the value of the ground set rank function. To see this, we first prove a useful lemma.

9.3.2. Lemma. *Let* $f_1, f_2 : 2^E \rightarrow \mathbb{R}$ *be such that* f_1 *is submodular and* f_2 *is supermodular, i.e., for* $A, B \subseteq E$,

$$f_2(A \cup B) + f_2(A \cap B) \geqslant f_2(A) + f_2(B).$$

Then if $f_1 \geqslant f_2$,

$$\mathscr{D}(f_1, f_2) = \{ A \subseteq E : f_1(A) = f_2(A) \}$$

is closed under union and intersection.

Proof. Let $A, B \in \mathscr{D}(f_1, f_2)$. Then

$$
\begin{aligned}
f_1(A) + f_1(B) &\geqslant f_1(A \cup B) + f_1(A \cap B) \\
&\geqslant f_2(A \cup B) + f_2(A \cap B) \\
&\geqslant f_2(A) + f_2(B)
\end{aligned}
$$

implies equality.   □

We apply the lemma to show that the integral matroid $Q(f)$ determines its ground set rank function.

9.3.3. Proposition. *Let* $Q(f)$ *be an integral matroid on* E *with ground set rank function* f. *Then for* $A \subseteq E$,

$$f(A) = \max \{ x(A) : x \in Q(f) \}.$$

Proof. By induction on $|E|$, we may assume $A = E$ and choose $x \in Q(f)$ so that $x(E)$ is maximal.

Set $\mathscr{D} = \{ S \subseteq E : x(S) = f(S) \}$ and $D = \bigcup \{ S : S \in \mathscr{D} \}$.

If $D = E$, then Lemma 9.3.2 implies $x(E) = f(E)$ because $S \rightarrow x(S)$ is a supermodular function.

If $z \in E \setminus D$, then $x(S) < f(S)$ for all $S \subseteq E$ with $z \in S$. Hence $x + z \in Q(f)$ contradicts the maximality of x, where in the notation '$x + z$' the element z is identified with its characteristic vector on E.   □

9.3.4. Corollary. *If* $c \in \mathbb{N}^E$ *is a bounding vector for* $Q(f)$, *then for all* $A \subseteq E$, $f(A) = r(C(A))$.   □

Let us further clarify the relation between the integral matroid $Q(f)$ and its Dilworth completion $M(f; c)$ on $P(c)$. It is convenient to introduce an operation that 'pushes down' a set $S \subseteq P(c)$ as much as possible to turn it into an order ideal of the same cardinality. Thus for $S \subseteq P(c)$, we let id(S) be the unique ideal in $P(c)$ such that

$$|\text{id}(S) \cap C(e)| = |S \cap C(e)| \quad \text{for every} \quad e \in E.$$

The vector associated with id(S) is the vector $x \in Q(f)$ so that $x(e) = |S \cap C(e)|$ for all $e \in E$.

9.3.5. Proposition. *Let M $(f;c)$ be the Dilworth embedding of the integral matroid $Q(f)$. Then for every $S \subseteq E$,*

$$r(S) = r(\mathrm{id}\,(S)).$$ $\qquad\square$

This proposition now allows us to immediately carry over the structural results derived for matroids in Section 9.2 to integral matroids. In particular, Corollary 9.2.9 yields the *intersection theorem for integral matroids*:

9.3.6. Corollary. *Let $Q(f_1)$ and $Q(f_2)$ be two integral matroids in $\mathbb{N}^E$. Then*

$$max\,\{x(E) : x \in Q(f_1) \cap Q(f_2)\} = min\,\{f_1(A) + f_2(E \backslash A) : A \subseteq E\}.$$

Proof. By Proposition 9.3.5, we may compute the left-hand side in the Dilworth embeddings $M(f_1;c)$ and $M(f_2;c)$, where c is chosen so as to bound both $Q(f_1)$ and $Q(f_2)$.

By Corollary 9.2.9, there exists $S \subseteq P(c)$ so that

$$r_1(S) + r_2(P(c) \backslash S)$$

attains the optimal value of the left-hand side. Let $A \subseteq E$ be such that

$$r_1(S) = f_1(A) + |S \backslash C(A)|.$$

Then we can assume $S \supseteq C(A)$ since we want to minimize the right-hand side. Thus

$$f_1(A) = r_1(S) - |S \backslash C(A)| \quad \text{and}$$
$$f_2(E \backslash A) \leqslant r_2(P(c) \backslash S) + |S \backslash C(A)|.$$

Hence

$$f_1(A) + f_2(E \backslash A) \leqslant \max\,\{x(E) : x \in Q(f_1) \cap Q(f_2)\}$$
$$\leqslant f_1(A) + f_2(E \backslash A). \qquad\square$$

A similar argument applies to the weighted case. Let $w : E \to \mathbb{R}$ be a weighting. Then w extends to $\bar{w} : P(c) \to \mathbb{R}$ via, for $s \in P(s)$,

$$\bar{w}(s) = w(e) \quad \text{if and only if} \quad s \in C(e).$$

With respect to the Dilworth embedding $M(f;c)$, consider a chain

$$\varnothing = S_0 \subseteq S_1 \subseteq \cdots \subseteq S_k \subseteq S_{k+1} = P(c)$$

of subsets of $P(c)$ and a vector $\bar{y} \geqslant 0$ such that

$$\bar{y}(S) \neq 0 \quad \text{implies} \quad S = S_i \quad \text{for some} \quad i = 1, \ldots, k \qquad (9.20)$$

and

$$\sum \{\bar{y}(S) : t \in S, S \subseteq P(c)\} \geqslant \bar{w}(t) \quad \text{for all } t \in P(c). \qquad (9.21)$$

We want to minimize

$$\sum_{S \subseteq P(c)} y(S) r_1(S) + \sum_{S \subseteq P(c)} y(S) r_2(S)$$

respecting conditions (9.20) and (9.21).

Suppose there exist some $e \in E$ and $t \in P(c)$ so that $t \notin S_k$. Then the chain $C(e)$ is not needed in order to satisfy (9.20) and (9.21), that is, we may assume for all $e \in E$, either $C(e) \subseteq S_k$ or $C(e) \cap S_k = \varnothing$. Thus S_k can be assumed to be of the form $S_k = C(A)$ for some $A \subseteq E$. Similarly, if $t \in C(e)$ is such that $t \in S_i$ and

$$y(S_i) + y(S_{i+1}) + \cdots + y(S_{k+1}) \geqslant \bar{w}(t)$$

and

$$y(S_{i+1}) + \cdots + y(S_{k+1}) < \bar{w}(t),$$

then $C(e) \subseteq S_i$ must hold and $S_{i-1} \cap C(e) = \varnothing$ can be assumed for (9.20) and (9.21) to hold. Hence all the S_i's may be chosen to be of the form $C(A)$.

Whence we obtain a generalization of Theorem 9.2.8 to the case of two ground set rank functions f_1 and f_2 and

$$\mathbb{P}(f_1) = \{x \in \mathbb{R}^E : x \geqslant 0, x(A) \leqslant f(A) \quad \text{for} \quad A \subseteq E\},$$
$$\mathbb{P}(f_2) = \{x \in \mathbb{R}^E : x \geqslant 0, x(A) \leqslant f(A) \quad \text{for} \quad A \subseteq E\}.$$

9.3.7. Proposition. *The linear program*

$$max \ w \cdot x \quad subject \ to \quad x \in \mathbb{P}(f_1) \cap \mathbb{P}(f_2) \tag{9.22}$$

is totally dual integral.

Proof. We compute the optimal solution with respect to $Q(f_1) \cap Q(f_2)$ in the matroids $M(f_1; c)$ and $M(f_2; c)$.

As the preceding discussion shows, the dual optimal solutions with respect to $M(f_1; c)$ and $M(f_2; c)$ can be interpreted to be optimal dual solutions for the LP-dual of (9.22):

$$\min \sum_{A \subseteq E} y_1(A) f_1(A) + \sum_{A \subseteq E} y_2(A) f_2(A)$$

subject to $y_1, y \geqslant 0$ and for all $e \in E$,

$$\sum \{y_1(A) + y_2(A) : e \in A, A \subseteq E\} \geqslant w(e). \qquad \square$$

As a special case we obtain a solution for the linear optimization problem over one integral matroid $Q(f)$:

$$max \ w \cdot x \quad subject \ to \quad x \in Q(f) \tag{9.23}$$

The greedy algorithm for (9.23) generalizes the greedy algorithm 9.1.4.

9.3.8. Greedy Algorithm.

(i) *Order the elements of E so that*

$$w(e_1) \geqslant w(e_2) \geqslant \cdots \geqslant w(e_k) > 0 \geqslant w(e_{k+1}) \geqslant \cdots \geqslant w(e_n);$$

(ii) $A \leftarrow \varnothing;\ i \leftarrow 1$;
(iii) IF $i > k$ THEN GOTO (vii);
(iv) $x(i) \leftarrow f(A \cup e_i) - f(A)$;
(v) $A \leftarrow A \cup e_i;\ i \leftarrow i + 1$;
(vi) GOTO (iii);
(vii) $x(i) \leftarrow 0$;
(viii) $i \leftarrow i + 1$;
(ix) IF $i \leqslant n$ THEN GOTO (vii);
(x) STOP.

9.3.9. Corollary. *The greedy algorithm 9.3.8 is optimal for the optimization problem (9.23).*   □

Also the greedy algorithm 9.3.8, of course, can be interpreted within the Dilworth embedding $M(f; c)$ of the integral matroid $Q(f)$ (see Exercise 9.9). In fact, a closer look at our analysis of the 'primal' optimization problems considered so far reveals that the structure of integral matroids relies mainly on the ordinal but not so much on the cardinal properties of the natural numbers. From this point of view, integral matroids may be seen as instances of 'supermatroids' (cf. Dunstan, Ingleton, and Welsh 1972, Welsh 1976) as well as 'matroids on (partially) ordered sets' (cf. Faigle 1980). Whereas the former abstract the notion of independence systems, the latter offer a theory of general semimodular closure operators. The greedy algorithm 9.3.8 can, in this context, be exhibited as a special case of a more general greedy algorithm for ordered sets (cf. Faigle 1979, 1985).

The matroid-sum construction was introduced in the previous section. It presents itself as a geometrically more natural notion in the context of integral matroids. As before, sums and intersections of two integral matroids are, in a sense, dual notions. So, let us first study the notion of a (Whitney) dual of an integral matroid $Q(f)$ with Dilworth embedding $M(f; c)$.

Formally, given the ground set rank function f, we define for every $A \subseteq E$,

$$f^*(A) = c(A) + f(E \setminus A) - f(E) \tag{9.24}$$

and call f^* the *c-dual* of f (McDiarmid 1975). Apparently, f^* also is a ground set rank function of an integral matroid bounded by c (Exercise 9.10).

9.3.10. Proposition. *If f^* is the c-dual of the ground set rank function f, then $M^* = M(f^*; c)$ is the Whitney dual of $M = M(f; c)$.*

Proof. Denoting by r^* the rank function of M^*, we have

$$r^*(S) = \min \{ f^*(A) + |S \backslash (A)| : A \subseteq E \}$$
$$= \min \{ f(E \backslash A) + |S| - |S \cap C(A)| + |C(A)| : A \subseteq E \} - f(E)$$
$$= |S| + \min \{ f(E \backslash A) + |(P(c) \backslash S) \backslash C(E \backslash A)| : A \subseteq E \} - f(E)$$
$$= |S| + r(P(c) \backslash S) - r(P(c)).\qquad\qquad \Box$$

Definition (9.24) yields a very strong dependence on the particular choice of the bounding vector c. In our investigation until now, this did not matter so much. The bounding vector c just offered a convenient way to reduce integral matroid theory to 'classical' matroid theory. This dependence can be dispensed with as follows.

With the ground set rank function f we associate its *supermodular* (!) *dual* $f^\#$, where for all $A \subseteq E$,

$$f^\#(A) = f(E) - f(E \backslash A),$$

and the unbounded *supermodular system*

$$Q^\#(f^\#) = \{ x \in \mathbb{Z}^E : x(A) \geqslant f^\#(A) \quad \text{for all} \quad A \subseteq E \}.$$

Purely formally, $(f^\#)^\# = f$ and for the c-dual f^* of f, $f^* = c - f$.

9.3.11. Proposition. *Let* $B(f) = \{ x \in Q(f) : x(E) = f(E) \}$ *be the set of bases of* $Q(f)$. *Then*

$$B(f) = Q(f) \cap Q^\#(f^\#).$$

Moreover, for every $A \subseteq E$,

$$f^\#(A) = \min \{ x(A) : x \in Q^\#(f^\#) \}.$$

Proof. Exercise 9.11. $\qquad\qquad\qquad\qquad\qquad\qquad\qquad\qquad\qquad \Box$

Thus, if we set $Q^*(f) = \{ x \in Q^\#(f^\#) : x \leqslant c \}$, then $Q^*(f)$ essentially describes the integral matroid $Q(f^*)$ associated with the c-dual f^* of f.

Let us now return to the sum of two integral matroids $Q(f_1)$ and $Q(f_2)$. Observing that $f = f_1 + f_2$ again is a ground set rank function, we call the integral matroid $Q(f_1 + f_2)$ the *sum* of $Q(f_1)$ and $Q(f_2)$. The *sum theorem for integral matroids* then says:

9.3.12. Theorem. $Q(f_1 + f_2)$ *is the vector sum of* $Q(f_1)$ *and* $Q(f_2)$, *i.e.,*

$$Q(f_1 + f_2) = Q(f_1) + Q(f_2) = \{ x_1 + x_2 : x_1 \in Q(f_1), x_2 \in Q(f_2) \}.$$

Proof. Since $Q(f_1) + Q(f_2) \subseteq Q(f_1 + f_2)$ is clear, suppose there exists a vector $b \in Q(f_1 + f_2)$ such that $b \neq x_1 + x_2$ whenever $x_1 \in Q(f_1)$ and $x_2 \in Q(f_2)$. Without loss of generality, we may assume that b is a basis of $Q(b_1 + b_2)$, i.e., $b(E) = f_1(E) + f_2(E)$.

Consider the ground set rank function f'_1, where

$$f'_1(S) = \min\{f_1(X) + |S\backslash X| : X \subseteq E\}$$

(cf. Exercise 9.11). We claim $b \in Q(f'_1 + f_2)$. If not, we must have

$$b(A) > f'_1(A) + f_2(A)$$

for some $A \subseteq E$. We now can find a set $Z \subseteq A$ so that $f'_1(A) = f_1(Z) + |A\backslash Z|$. But then

$$b(Z) = b(A) - b(A\backslash Z) > f_1(Z) + f_2(A) \geqslant f_1(Z) + f_2(Z)$$

contradicts the choice of b.

Thus $f'_1(E) = f_1(E)$. Moreover, since $f'_1 \leqslant f_1$, we may assume that b bounds $Q(f_1)$ and $Q(f_2)$ (otherwise, we replace f_1 and f_2 by f'_1 and f'_2). Let $Q(f_2^*)$ be the b-dual of $Q(f_2)$ and apply Corollary 9.3.6 to $Q(f_1)$ and $Q(f_2^*)$. For every $A \subseteq E$,

$$f_1(A) + f_2^*(E\backslash A) = f_1(A) + b(E\backslash A) + f_2(A) - f_2(E) \geqslant f_1(E).$$

Therefore, there exists $x_1 \in Q(f_1) \cap Q(f_2^*)$ with $x_1(E) = f_1(E)$. Hence $b(E) = f_1(E) + f_2(E)$ implies $x_2 = b - x_1 \in Q(f_2)$, contradicting the choice of b. $\qquad\square$

A direct application of the sum theorem yields the integral *separation theorem* for sub- and supermodular functions of Frank (1982):

9.3.13. Corollary. *Let* $f, g : 2^E \rightarrow \mathbb{Z}$ *be functions such that* $g \leqslant f, g(\varnothing) = 0 = f(\varnothing), f$ *is sub- and* g *is supermodular. Then there exists a function* $m : 2^E \rightarrow \mathbb{Z}$ *which is modular, i.e., both sub- and supermodular, and satisfies* $g \leqslant m \leqslant f$.

Proof. Observe that $(-g)$ is submodular and choose the vector $v \in \mathbb{N}^E$ with large enough components so that $f + v$ and $(-g) + v$ are ground set rank functions.

Because $f - g \geqslant 0$, we have for all $A \subseteq E$,

$$2v(A) \leqslant (f + v)(A) + (-g + v)(A).$$

Hence, there are $y_1 \in Q(f + v)$ and $y_2 \in Q(-g + v)$ with

$$2v = y_1 + y_2.$$

Now

$$0 = (y_1 - v) + (y_2 - v)$$

yields

$$y_1 - v = -(y_2 - v)$$

and, therefore, $m = y_1 - v \in \mathbb{Z}^E$ satisfies, for all $A \subseteq E$,

$$g(A) \leqslant m(A) \leqslant f(A). \qquad \square$$

Note that the matroid union $M_1 \vee M_2$, which we considered previously, can also be cast into the framework of sums of integral matroids: we represent M_1 and M_2 by the collections Q_1 and Q_2 of characteristic vectors of their respective independent sets. The independent sets of $M_1 \vee M_2$ then correspond to the restriction of $Q_1 + Q_2$ to

$$D(1) = \{x \in \mathbb{N}^E : x(e) \leqslant 1 \quad \text{for all} \quad e \in E\}.$$

In summary, we have seen that integral matroids allow a structural matroid-theoretic analysis *via* their Dilworth embeddings. In this sense, all structural properties are just ordered versions of matroid properties.

In particular, our trans-shipment problem at the beginning of this section is a matroid intersection problem. If $G = (S, T; E)$ is the bipartite graph with supply vector $a \in \mathbb{N}^S$ and demand vector $d \in \mathbb{N}^T$, we define two integral matroids on the set E of edges:

$$Q_S = \{x \in \mathbb{N}^E : x(U) \leqslant a(U) \quad \text{for all} \quad U \subseteq S\},$$
$$Q_T = \{x \in \mathbb{N}^E : x(V) \leqslant d(V) \quad \text{for all} \quad V \subseteq T\}.$$

It is not difficult to see that Q_S and Q_T are indeed integral matroids. Thus the demand can be satisfied if and only if

$$\max \{x(E) : x \in Q_S \cap Q_T\} = d(T).$$

A capacity restriction $c \in \mathbb{N}^E$ on E can be taken care of similarly by restricting Q_S and Q_T suitably.

The question now arises whether an even more general extension of matroid theory exists. For example, do submodular functions fit into this context when they are not necessarily integer valued or non-decreasing?

The answer is yes. The linear programming approach of Edmonds (1970) makes no general integrality assumption. On the other hand, rational-valued submodular functions are essentially integral. We only have to multiply the (finitely many) rationals by a suitable large integer. The general case of real numbers can then be handled by straightforward rational approximation (cf. McDiarmid 1975).

The separation theorem (Corollary 9.3.13) already indicates that monotonicity of the sub- and supermodular functions in question is not really essential. Moreover, the discussion of integral matroid duality hints at the possibility of including systems which need not be bounded (or non-negative, for that matter). Furthermore, the ground set rank function need not be defined for all subsets of E. We will turn our attention to such systems in the next section.

9.4. Submodular Systems

The generalizing step from matroids to integral matroids consisted in allowing monotone submodular ground set rank functions which no longer need to have the unit increase property. We will now relax the axiomatic requirements our 'ground set rank functions' should satisfy even further and thus broaden the scope of the theory. Yet, we will still be able to reduce structural questions to questions about matroids.

Rather than considering all subsets of the ground set E, we assume we are given a collection $\mathscr{D} \subseteq 2^E$ of subsets such that

$$\varnothing \in \mathscr{D} \quad \text{and} \quad E \in \mathscr{D}, \tag{9.25}$$

$$A \cup B \in \mathscr{D} \quad \text{and} \quad A \cap B \in \mathscr{D} \quad \text{for all} \quad A, B \in \mathscr{D}. \tag{9.26}$$

We say that the integer-valued function $f: \mathscr{D} \to \mathbb{Z}$ is *normalized submodular* on $\mathscr{D}$ if

$$f(\varnothing) = 0, \tag{9.27}$$

$$f(A \cup B) + f(A \cap B) \leqslant f(A) + f(B) \quad \text{for all} \quad A, B \in \mathscr{D}. \tag{9.28}$$

Following Fujishige (1984a), we call $(\mathscr{D}, f)$ a *submodular system* if (9.25)–(9.28) hold. With the submodular system $(\mathscr{D}, f)$ we associate the *submodular structure*

$$S(\mathscr{D}, f) = \{x \in \mathbb{Z}^E : x(A) \leqslant f(A) \quad \text{for all} \quad A \in \mathscr{D}\}.$$

Thus submodular structures are unbounded analogs of integral matroids.

In the same way, $(\mathscr{D}, g)$ is a *supermodular system* if the function $g: \mathscr{D} \to \mathbb{Z}$ satisfies

$$g(g\varnothing) = 0,$$

$$g(A \cup B) + g(A \cap B) \geqslant g(A) + g(B) \quad \text{for all} \quad A, B \in \mathscr{D},$$

and thus is *normalized supermodular*. The associated *supermodular structure* is defined as

$$S^{\#}(\mathscr{D}, g) = \{x \in \mathbb{Z}^E : x(A) \geqslant g(A) \quad \text{for all} \quad A \in \mathscr{D}\}.$$

The supermodular system $(\mathscr{D}', g)$ is *dual* to the submodular system $(\mathscr{D}, f)$ if

$$\mathscr{D}' = \mathscr{D}^* = \{A \subseteq E : E \setminus A \in \mathscr{D}\},$$

$$g(A) = f^{\#}(A) = f(E) \setminus f(E \setminus A) \quad \text{for all} \quad A \in \mathscr{D}^*.$$

The next relation follows directly from the definition of sub- and supermodular structures:

$$\mathbb{B}(\mathscr{D}, f) = \{x \in S(\mathscr{D}, f) : x(E) = f(E)\}$$
$$= \{x \in S^{\#}(\mathscr{D}^{\#}, f^{\#}) : x(E) = f^{\#}(E)\}$$
$$= S(\mathscr{D}, f) \cap S^{\#}(\mathscr{D}^{\#}, f^{\#}).$$

$\mathbb{B}(\mathscr{D}, f)$ is the basis structure of $(\mathscr{D}, f)$ and $(\mathscr{D}^{\#}, f^{\#})$ respectively. We will see below that it determines both $\mathscr{D}$ and f uniquely. For this purpose, we associate with the submodular system $(\mathscr{D}, f)$ an auxiliary integral matroid by a construction which we already have used in the proof of Corollary 9.3.13.

Choose a vector $v \in \mathbb{N}^E$ such that the submodular function $A \to f(A) + v(A)$ is strictly increasing on $\mathscr{D}$, i.e., for all $A, B \in \mathscr{D}$ with $A \neq B$,

$$A \subset B \quad \text{implies} \quad f(A) + v(A) < f(B) + v(B).$$

We now extend $f + v$ to a submodular function $\bar{f}_v$ defined for *all* subsets $S \subseteq E$ via

$$\bar{f}_v(S) = \min\{f(A) + v(A) : S \subseteq A, \quad A \in \mathscr{D}\}.$$

It is not difficult to verify that $\bar{f}_v$ is indeed a ground set rank function in the sense of Section 9.3. We are interested in the non-negative part of the translation $\mathbb{B}(f + v)$ of the basis structure $\mathbb{B}(f)$.

9.4.1. Theorem. *Under the conditions above,*

$$\{x \in \mathbb{B}(f + v) : x \geqslant 0\} = \{x \in Q(\bar{f}_v) : x(E) = f(E) + v(E)\}. \qquad \square$$

9.4.2. Corollary. *The submodular system $(\mathscr{D}, f)$ is uniquely determined by its basis structure $\mathbb{B}(f)$.*

Proof. Choosing the vector $v \in \mathbb{N}^E$ as in the hypothesis of Theorem 9.4.1, we note that $\mathbb{B}(f)$ determines the collection of bases of the integral matroid $Q(\bar{f}_v)$ and hence, by Proposition 9.3.3, the ground set rank function $\bar{f}_v$. Furthermore, by the construction of $\bar{f}_v$ from the strictly increasing function $f + v$, we have

$$\bar{f}_v(A) = f(A) + v(A) \quad \text{for all} \quad A \in \mathscr{D},$$

i.e.,

$$f(A) = \bar{f}_v(A) - v(A) \quad \text{for all} \quad A \in \mathscr{D},$$

and

$$\mathscr{D} = \{A \subseteq E : \bar{f}_v(A) < \bar{f}_v(A \cup x) \quad \text{for all} \quad x \in E \backslash A\}. \qquad \square$$

With the same method of proof, we obtain the *sum theorem* for submodular systems:

9.4.3. Corollary. *Let $(\mathscr{D}_1, f_1)$ and $(\mathscr{D}_2, f_2)$ be submodular systems and consider the submodular system $(\mathscr{D}, f)$, where $\mathscr{D} = \mathscr{D}_1 \cap \mathscr{D}_2$ and $f = f_1 + f_2$. Then the associated submodular structures satisfy*

$$S(\mathscr{D}, f) = S(\mathscr{D}_1, f_1) + S(\mathscr{D}_2, f_2)$$
$$= \{x + y : x \in S(\mathscr{D}_1, f_1), \quad y \in S(\mathscr{D}_2, f_2)\}.$$

Proof. Exercise 9.13. $\qquad \square$

The *intersection theorem* for submodular systems, which was seen to imply the sum theorem in the last section, can also be regarded as a consequence of the sum theorem:

9.4.4. Corollary. *Let $(\mathscr{D}_1, f_2)$ and $(\mathscr{D}_2, f_2)$ be submodular systems and $k\in\mathbb{Z}$ a fixed integer. Then there exists $x\in S(\mathscr{D}_1, f_1)\cap S(\mathscr{D}_2, f_2)$ with $x(E)\geqslant k$ if and only if for all $A\in\mathscr{D}_1$ so that $(E\backslash A)\in\mathscr{D}_2$,*

$$f_1(A) + f_2(E\backslash A) \geqslant k. \tag{9.29}$$

Proof. Condition (9.29) is obviously necessary. Let us show sufficiency.

We may assume that both f_1 and f_2 are non-decreasing on $\mathscr{D}_1$ and $\mathscr{D}_2$ [otherwise, we add a suitable $v\in\mathbb{N}^E$ to f_1 and f_2 and replace k by $k + v(E)$]. Replacing $f_i(i = 1, 2)$ by $\min\{f_1, k\}$ if necessary, we furthermore assume

$$f_1(E) = f_2(E) = k.$$

If $\mathscr{D}_2^*$ is now the family of set-theoretic complements of $\mathscr{D}_2$, we conclude from (9.29) for all $A\in\mathscr{D}_1\cap\mathscr{D}_2^*$,

$$0 \leqslant f_1(A) - f_2^\#(A).$$

By Corollary 9.4.3, we can thus find $x\in S(D_1, f_1)$ and $y\in S(\mathscr{D}_2^*, -f_2)$ so that $0 = x + y$.

Because $f_1(E) - f_2^\#(E) = 0$, the vector 0 must be a basis vector of $S(\mathscr{D}_1\cap D_2^*, f_1 - f_2^\#)$. Hence we must have $x(E) = f_1(E) = k$. Moreover, for every $B\in\mathscr{D}_2$,

$$x(B) = x(E) + y(E\backslash B) \leqslant k - f_2^\#(E\backslash B) \leqslant f_2(B),$$

i.e., x is also a member of $S(\mathscr{D}_2, f_2)$ with the desired properties. $\qquad\square$

Let us turn our attention to the optimization problem

$$\max w\cdot x \quad \text{subject to} \quad x\in\mathbb{B}(f), \tag{9.30}$$

where $\mathbb{B}(f)$ is the basis structure of the submodular system $(\mathscr{D}, f)$ and $w: E\to\mathbb{R}$ is a weight vector. We assume that w is $\mathscr{D}$-*compatible*, i.e., for every $c\in\mathbb{R}$,

$$\{x\in E: w(x)\geqslant c\}\in\mathscr{D}, \tag{9.31}$$

since the problem (9.30) is unbounded otherwise (cf. Exercise 9.14). We extend the greedy algorithm 9.3.3 and define a *greedy solution* for (9.30) to be a vector $x^*\in\mathbb{Z}^E$ obtained by the following construction.

Let $\varnothing = D_0\subset D_1\subset D_2\subset\cdots\subset D_k = E$ be a maximal chain of sets in $\mathscr{D}$ containing all the sets of the form (9.31). Choose k elements $e_i\in D_i\backslash D_{i-1}$, $i = 1,\ldots,k$, and set

$$x^*(e) = \begin{cases} f(D_i) - f(D_{i-1}) & \text{if } e = e_i, 1\leqslant i\leqslant k, \\ 0 & \text{otherwise.} \end{cases}$$

9.4.5. Proposition. *If $w: E \to \mathbb{R}$ is $\mathscr{D}$-compatible, then the greedy solution x^* is optimal for the optimization problem (9.30).*

Proof. Choose $v \in \mathbb{N}^E$ as in the proof of Corollary 9.4.2 and observe that the vector $x^* + v$ is in compliance with the greedy algorithm 9.3.8 with respect to $\tilde{f}_v$ and, therefore, optimal for the optimization problem

$$\max w \cdot y \quad \text{subject to} \quad y \in \{b \in \mathbb{B}(f + v) : b \geqslant 0\}.$$

Suppose there exists $x' \in \mathbb{B}(f)$ with $w \cdot x' > x^*$. Then, choosing v in addition as to satisfy $(x' + v) \geqslant 0$, we have

$$(x' + v) \in \{b \in \mathbb{B}(f + v) : b \geqslant 0\} \quad \text{and} \quad w \cdot (x' + v) > w \cdot (x^* + v),$$

which is contradiction. $\qquad\qquad\qquad\qquad\qquad\qquad\qquad\qquad\qquad\qquad\qquad\square$

A further generalization of integral matroids is due to Frank (1981). We consider a submodular system $(\mathscr{D}_1, f)$ and a supermodular system $(\mathscr{D}_2, g)$ such that for all $A \in \mathscr{D}_1$ and $B \in \mathscr{D}_2$,

$$A \backslash B \in \mathscr{D}_1 \quad \text{and} \quad B \backslash A \in \mathscr{D}_2, \tag{9.32}$$

$$f(A) - g(B) \geqslant f(A \backslash B) - g(B \backslash A). \tag{9.33}$$

Then

$$S(\mathscr{D}_1, f; \mathscr{D}_2, g) = S(\mathscr{D}_1, f) \cap S^{\#}(\mathscr{D}_2, g)$$

is the *generalized integral matroid* determined by $(\mathscr{D}_1, f)$ and $(\mathscr{D}_2, g)$. (We remark that this definition can be given, seemingly more generally, in terms of the 'intersecting families' below.)

With respect to $S(\mathscr{D}_1, f; \mathscr{D}_2, g)$ let us adjoin a new element $\hat{e}$ to our ground set E and set

$$\hat{E} = E \cup \{\hat{e}\}.$$

We furthermore define

$$\mathscr{D}_2^* = \{\hat{E} \backslash B : B \in \mathscr{D}_2\},$$

$$\hat{\mathscr{D}} = \mathscr{D}_1 \cup \mathscr{D}_2^*$$

and note that $\hat{\mathscr{D}}$ is closed under taking unions and intersections and contains the new ground set $\hat{E}$ as a member.

Fixing an arbitrary integer $m \in \mathbb{Z}$, we define $\hat{f} : \hat{\mathscr{D}} \to \mathbb{Z}$ by

$$\hat{f}(X) = \begin{cases} m & \text{if } X = \hat{E}, \\ f(X) & \text{if } X \in \mathscr{D}_1, \\ \hat{f}(\hat{E}) - g(E \backslash X) & \text{if } X \in \mathscr{D}_2^*. \end{cases}$$

To see that $\hat{f}$ is submodular on $\hat{\mathscr{D}}$, consider $X \in \mathscr{D}_1$ and $Y \in \mathscr{D}_2^*$. Then (9.32) and (9.33) imply

$$f(X \cap Y) = f(X \backslash (E \backslash Y)) \leqslant f(X) + g((E \backslash Y) \backslash X) - g(E \backslash Y),$$

and, therefore,

$$\hat{f}(X \cup Y) + \hat{f}(X \cap Y) = m - g(E \setminus (X \cup Y)) + f(X \cap Y), \leqslant \hat{f}(X) + \hat{f}(Y).$$

The result of Schrijver (1982) and Fujishige (1984b) reduces generalized matroid theory to the study of basis structures:

9.4.6. Proposition. $S(\mathcal{D}_1, f; \mathcal{D}_2, g)$ *is the projection of the basis structure* $\mathbb{B}(\hat{f})$ *along* $\hat{e}$, *i.e.*,

$$S(\mathcal{D}_1, f; \mathcal{D}_2, g) = \{x \in \mathbb{Z}^E: \quad \text{there exists} \quad \hat{x} \in \mathbb{B}(\hat{f})$$
$$\text{so that} \quad \hat{x}(e) = x(e) \quad \text{for all} \quad e \in E\}.$$

Proof. Assume $\hat{x} \in \mathbb{B}(\hat{f})$. Then $A \in \mathcal{D}_1$ implies

$$\hat{x}(A) \leqslant \hat{f}(A) = f(A),$$

and $B \in \mathcal{D}_2$ implies

$$\hat{x}(B) = \hat{x}(E) - \hat{x}(\bar{E} \setminus B) \geqslant \hat{f}(\bar{E}) - \hat{f}(\bar{E}) + g(B) = g(B).$$

Conversely, if $x \in S(\mathcal{D}_1, f; \mathcal{D}_2, g)$ we define $\hat{x} \in \mathbb{Z}^E$ by

$$\hat{x}(e) = \begin{cases} x(e) & \text{if } e \in E, \\ \hat{f}(E) - x(E) & \text{if } e = \hat{e}. \end{cases}$$

To verify $\hat{x} \in \mathbb{B}(\hat{f})$, we just note for $\hat{E} \setminus B \in \hat{\mathcal{D}}_2^*$,

$$\hat{x}(\hat{E} \setminus B) = \hat{x}(\hat{E}) - x(B) \leqslant \hat{f}(E) - g(B). \qquad \square$$

A converse of Proposition 9.4.6 can be obtained as follows. Distinguish an arbitrary element $\hat{e} \in E$ and set $E' = E \setminus \{\hat{e}\}$. Then associate with the submodular system $(\mathcal{D}, f)$ on E the submodular system $(\mathcal{D}_1, f')$ on E' and the supermodular system (D_2, g) on E', where

$$\mathcal{D}_1 = \{A \in \mathcal{D} : \hat{e} \notin A\},$$
$$\mathcal{D}_2 = \{E \setminus X : \hat{e} \in X, X \in \mathcal{D}\},$$
$$f'(A) = f(A) \quad \text{for all} \quad A \in \mathcal{D}_1,$$
$$g(B) = f^{\#}(B) = f(E) - f(E \setminus B) \quad \text{for all} \quad B \in \mathcal{D}_2.$$

$S(\mathcal{D}_1, f'; \mathcal{D}_2, g)$ is then the projection of $\mathbb{B}(f)$ along $\hat{e}$ (cf. Exercise 9.15).

The *generalized matroids* of Tardos (1983) comprise the case where $\mathcal{D}_1$ and $\mathcal{D}_2$ contain all subsets of E and r_1 and r_2 are rank functions of matroids on E (in the 'classical' sense). Since we can identify subsets of E with characteristic vectors, we thus consider

$$M(r_1, r_2^{\#}) = \{A \subseteq E : r_2(E) - r_2(E \setminus X) \leqslant |A \cap X| \leqslant r_1(X) \quad \text{for all} \quad X \subseteq E\}.$$

Observe that each set $A \in M(r_1, r_2^{\#})$ not only is independent in the matroid $M(r_1)$ but also satisfies $r_2(A) = r_2(E)$, i.e., is spanning in the matroid $M(r_2)$. In other words, generalized matroids are intersections of the collections of independent and of spanning sets of pairs of matroids.

The construction of Proposition 9.4.6 allows us to represent $M(r_1, r_2^{\#})$ as the projection of the collection of bases of some integral matroid $Q(\hat{f})$. Embedding $Q(\hat{f})$ into its Dilworth completion with respect to a vector $c \in \mathbb{N}^E$ such that $c(e) \leq 1$ for all $e \in E$, essentially yields

9.4.7. Corollary. *A non-empty collection $\mathscr{I}$ of subsets of E is a generalized matroid if and only if $\mathscr{I}$ is the projection onto E of the collection of bases of a matroid defined on $E \cup T$, where the set T is disjoint from E.*   □

So far, the sub- and supermodular functions we have considered as generalized ground set rank functions were defined on families of subsets which are closed under union and intersection. This model can be relaxed even further.

We say that the two subsets $A, B \subseteq E$ are *intersecting* if $A \cap B \neq \varnothing$. An *intersecting family* is a non-empty collection $\mathscr{K}$ of subsets of E such that $\varnothing \notin \mathscr{K}$ and for all intersecting members $A, B \in \mathscr{K}$,

$$A \cap B \in \mathscr{K} \quad \text{and} \quad A \cup B \in \mathscr{K}.$$

The function $f: \mathscr{K} \to \mathbb{Z}$ is *submodular* on the intersecting family $\mathscr{K}$ if for all intersecting members $A, B \in \mathscr{K}$,

$$f(A \cup B) + f(A \cap B) \leq f(A) + f(B).$$

Similarly, $g: \mathscr{K} \to \mathbb{Z}$ is *supermodular* on $\mathscr{K}$ if $(-g)$ is submodular. The key for the analysis of sub- and supermodular functions on intersecting families is provided by the next two fundamental observations:

9.4.8. Lemma. *Let $\mathscr{K}$ be an arbitrary family of subsets of E and $f: \mathscr{K} \to \mathbb{Z}$ a function. Then*

$$\{x \in \mathbb{Z}^E : x(A) \leq f(A) \text{ for all } A \in \mathscr{K}\} = \{x \in \mathbb{Z}^E : x(A) \leq \bar{f}(A) \text{ for all } A \in \bar{\mathscr{K}}\},$$

where $\bar{\mathscr{K}}$ is the collection of all subsets of E which can be partitioned into members of $\mathscr{K}$, i.e.,

$$\bar{\mathscr{K}} = \{X \subseteq E : X = X_1 \cup \cdots \cup X_m, X_i \in \mathscr{K}\} \cup \{\varnothing\} \tag{9.34}$$

and

$$\bar{f}(X) = \begin{cases} 0 & \text{if } X = \varnothing, \\ \min\{\sum f(X_i) : X = X_1 \cup \cdots \cup X_k, X_i \in \mathscr{K}\} & \text{otherwise.} \end{cases} \tag{9.35}$$

9.4.9. Lemma. *Let $\mathscr{K}$ be an intersecting family and $f: \mathscr{K} \to \mathbb{Z}$ submodular. Furthermore, let $\bar{\mathscr{K}}$ and $\bar{f}$ be defined as in (9.34) and (9.35). Then*

(a) $\bar{\mathscr{K}}$ is closed under union and intersection,

(b) $\bar{f}: \bar{\mathscr{K}} \to \mathbb{Z}$ is submodular.

Proof. Exercise 9.16.   □

It should be noted that $(\mathcal{K}, \bar{f})$ is not necessarily already a submodular system when $f: \mathcal{K} \to \mathbb{Z}$ is submodular on the intersecting family $\mathcal{K}$. The set E namely need not belong to $\mathcal{K}$. In this case, however, we may adjoin E to $\mathcal{K}$ and prescribe an arbitrary value $f_E(E)$ in order to obtain a submodular system $(\mathcal{K}_E, \bar{f}_E)$. An application of this principle yields

9.4.10. Proposition. *Let* $f: \mathcal{K} \to \mathbb{N}$ *and* $g: \mathcal{K} \to \mathbb{N}$ *be a sub- and a supermodular function on the intersecting family* $\mathcal{K}$ *and assume* $g(A) \leqslant |A|$ *for all* $A \in \mathcal{K}$. *Then*
  (a) $S(\mathcal{K}, f) = \{I \subseteq E : |I \cap A| \leqslant f(A)$ *for all* $A \in \mathcal{K}\}$ *is the collection of independent sets of a matroid on* E.
  (b) $S^*(\mathcal{K}, g) = \{S \subseteq E : |S \cup A| \geqslant g(A)$ *for all* $A \in \mathcal{K}\}$ *is the collection of spanning sets of a matroid on* E.

Proof. Without loss of generality, assume $E \in \mathcal{K}$. (Otherwise, adjoin E to $\mathcal{K}$ and set $g_E(E) = |E|$ and $f_E(E) = |E|$.)
    In view of Lemma 9.2.10, $\bar{f}: \mathcal{K} \to \mathbb{N}$ induces a matroid rank function on the subsets of E *via*

$$r(S) = \min\{\bar{f}(X) + |S \setminus X| : X \in \mathcal{K}\}.$$

Clearly, $r(S) \geqslant |S|$ if and only if $S \in S(\mathcal{K}, f)$. Hence (a) must hold.
    To verify (b), we consider the submodular function $A \to f(A) = |A| - g(A)$ on $\mathcal{K}$. By (a), f induces a matroid M on E. Now

$$|S \cap A| \geqslant g(A) \quad \text{iff} \quad |S \cap A| \geqslant |A| - f(A)$$
$$\text{iff} \quad f(A) \geqslant |A \setminus S| = |(E \setminus S) \cap A)|.$$

Thus the members of $S^{\#}(\mathcal{K}, g)$ are exactly the set-theoretic complements of the members of $S(\mathcal{K}, f)$ and hence are the spanning sets of the matroid dual M^*.
                                                                          □

The most general notion of a family of sets for which sub- and supermodularity of a function can be defined is due to Edmonds and Giles (1977).
    Two subsets $A, B \subseteq E$ form a *crossing pair* if $A \cap B \neq \emptyset$ and, dually, $A \cup B \neq E$. A collection $\mathcal{K}$ of subsets of E is a *crossing family* if $\emptyset, E \notin \mathcal{K}$ and for all crossing pairs, $A, B \in \mathcal{K}$

$$A \cup B \in \mathcal{K} \quad \text{and} \quad A \cap B \in \mathcal{K}.$$

$f: \mathcal{K} \to \mathbb{Z}$ is *submodular* on the crossing family $\mathcal{K}$ if for all crossing pairs $A, B \in \mathcal{K}$

$$f(A \cup B) + f(A \cap B) \leqslant f(A) + f(B). \tag{9.36}$$

$g: \mathcal{K} \to \mathbb{Z}$ is *supermodular* if $(-g)$ is submodular.
    This is a very weak notion of submodularity since property (9.36) is only

required to hold for pairs of subsets whose union does not cover the ground set E completely in case they intersect non-trivially at all. Let us look at an example to which we will come back in more detail in Section 9.5.

Let $N = (V, E; c)$ be a *network*, i.e., a directed graph with set V of nodes, set E of arcs, and a capacity function $c : E \to \mathbb{N}$. We consider the crossing family $\mathcal{K}$ on the set V of nodes of N consisting of all subsets of V except $\varnothing$ and E. For each $U \in \mathcal{K}$, we denote by $\hat{U}$ the *cut* determined by U, that is, the set of those arcs of N which have their initial node in U and their terminal node in $V \setminus U$. $c(U)$ is defined to be the sum of the capacities of the arcs in $\hat{U}$. Then $U \to c(U)$ is submodular on $\mathcal{K}$.

As Frank and Tardos (1982) have observed, the analogue of Proposition 9.4.10 is no longer true in the context of crossing families.

9.4.11. Example. *Let* $\mathcal{K} : \{\{a,b\}, \{b,c\}, \{a,c\}\}$ *be a crossing family on* $E = \{a,b,c\}$. *The function* $f : \mathcal{K} \to \mathbb{N}$ *with* $f(\{a,b\}) = 2$ *and* $f\{b,c\} = 1 = f\{a,c\}$ *is submodular on* $\mathcal{K}$. *Both* $\{a,b\}$ *and* $\{c\}$ *are maximal members of*

$$S(\mathcal{K}, f) = \{I \subseteq E : |I \cap A| \leqslant f(A) \quad for\ all\ A \in \mathcal{K}\}$$

and have different cardinality, which is impossible in matroid independence systems. □

In spite of the example, however, the theory of submodular functions on crossing families can be subsumed under the theory of integral matroids (and hence of matroids) in a manner similar to our treatment of submodular functions on intersecting families. Thus, let us fix a crossing family $\mathcal{K}$ on the ground set E and a submodular function $f : \mathcal{K} \to \mathbb{Z}$. We will complete $\mathcal{K}$ with respect to intersection and, dually, with respect to union and also use Lemma 9.4.9 with respect to submodularity and with respect to supermodularity in order to arrive at the desired equivalent basis structure.

As before, we associate with $\mathcal{K}$ the family

$$\bar{\mathcal{K}} = \{X \subseteq E : X \neq E, X = X_1 \cup \cdots \cup X_m, X_i \in \mathcal{K}\} \cup \{\varnothing\} \qquad (9.37)$$

of proper subsets of E which can be expressed as disjoint unions of members of $\mathcal{K}$. It is straightforward to verify for all $X, Y \in \bar{\mathcal{K}}$ such that $X \cup Y \neq E$,

$$X \cup Y \in \bar{\mathcal{K}} \quad and \quad X \cap Y \in \bar{\mathcal{K}}.$$

Passing to complements, we therefore obtain the intersecting family

$$(\bar{\mathcal{K}})^* = \{A \subseteq E : E \setminus A \in \bar{\mathcal{K}}\}.$$

9.4.12. Proposition. *Let* $f : \mathcal{K} \to \mathbb{Z}$ *be a submodular function on the crossing family* $\mathcal{K}$. *Then for every* $k \in \mathbb{Z}$,

$$\mathbb{B}_k(\mathcal{K}, f) = \{x \in \mathbb{Z}^E : x(E) = k,\ x(A) \leqslant f(A)\ for\ all\ A \in \mathcal{K}\}$$

is the basis structure of a (unique) submodular system.

Proof. Consider the crossing family

$$\mathcal{K}^* = \{E \backslash A : A \in \mathcal{K}\}$$

and the supermodular function $g : \mathcal{K}^* \to \mathbb{Z}$ given by

$$g(B) = k - f(E \backslash B)$$

for all $B \in \mathcal{K}^*$. Extend $\mathcal{K}^*$ to $\overline{(\mathcal{K}^*)}$ as in (9.37) and g to the supermodular function $\bar{g} : \overline{(\mathcal{K}^*)} \to \mathbb{Z}$ as in Lemma 9.4.9 i.e., for all $D \in \overline{(\mathcal{K}^*)}$,

$$\bar{g}(D) = \max \{\textstyle\sum g(D_i) : D = D_1 \cup \cdots \cup D_m, D_i \in \mathcal{K}^*\}.$$

Then

$$\mathcal{K}' = \overline{(\mathcal{K}^*)^*}$$

is an intersecting family and $f' : \mathcal{K}' \to \mathbb{Z}$, given for all $A \in \mathcal{K}'$ by

$$f'(A) = k - \bar{g}(E \backslash A)$$

is submodular on $\mathcal{K}'$.

Complete now $\mathcal{K}'$ to $\overline{\mathcal{K}'}$ and observe $E \in \overline{\mathcal{K}^*}$ because $\varnothing \in \overline{(\mathcal{K}')}$. $(\overline{\mathcal{K}'}, \overline{f'})$ is a submodular system with $\overline{f'}(E) = k$ and we have for every $x \in \mathbb{Z}^E$,

$$
\begin{array}{llll}
 & x(A) & \leqslant f(A) & \text{for all } A \in \mathcal{K} \\
\text{iff} & x(E \backslash A) & \geqslant k - f(A) & \text{for all } (E \backslash A) \in \mathcal{K}^* \\
\text{iff} & x(D) & \geqslant \bar{g}(D) & \text{for all } D \in \overline{(\mathcal{K}^*)} \\
\text{iff} & x(E \backslash D) & \leqslant f'(E \backslash D) & \text{for all } (E \backslash D) \in \mathcal{K}' \\
\text{iff} & x(B) & \leqslant f'(B) & \text{for all } B \in \overline{\mathcal{K}'}.
\end{array}
$$

Since the basis structure determines the associated submodular system uniquely (cf. Corollary 9.4.2), the proposition follows. $\qquad\square$

In this formulation, Proposition 9.4.12 is due to Fujishige (1984c). The matroidal version (Corollary 9.4.13 below) was proved by Frank and Tardos (1982).

Let us now briefly discuss how fundamental matroid operations are reflected in submodular structures and focus on the operations of restriction and contraction. $(\mathcal{D}, f)$ here is an arbitrary but fixed submodular system with submodular structure $S(\mathcal{D}, f)$.

For a given vector $c \in \mathbb{Z}^E$, we define the *vector rank*

$$r(c) = \max \{x(E) : x \in S(\mathcal{D}, f), x \leqslant c\}.$$

The intersection theorem (Corollary 9.4.4) then implies

$$r(c) = \min \{f(A) + c(E \backslash A) : A \in \mathcal{D}\}.$$

On the other hand, f induces a submodular function f^c on the Boolean algebra

$\mathscr{B}$ of *all* subsets of E *via*

$$f^c(S) = \min \{f(A) + c(S\backslash A): A\in\mathscr{D}\}$$

(cf. Exercise 9.11). In particular, $f^c(E) = r(c)$.

It is not hard to verify that the *restriction* of $S(\mathscr{D},f)$ to $\{x\in\mathbb{Z}^E: x \leqslant c\}$ is given by

$$\{x\in S(\mathscr{D},f): x \leqslant c\} = \{x\in\mathbb{Z}^E: x(S) \leqslant f^c(S) \text{ for all } S\subseteq E\}.$$

Dually, $\{x\in S(\mathscr{D},f): x \geqslant 0\}$ is either empty or an integral matroid (cf. Theorem 9.4.1). Hence for every $v\in S(\mathscr{D},f)$ the contraction of $S(\mathscr{D},f)$ by the vector v is the translation of an integral matroid by the vector v, i.e.,

$$\{x\in S(\mathscr{D},f): x \geqslant v\} = v + Q(f_v),$$

where $f_v(S) = \min \{f(A) - v(A): S\subseteq A, A\in\mathscr{D}\}$.

In other words, for every two vectors $v, c\in\mathbb{Z}^E$ such that $v\in S(\mathscr{D},f)$ and $v \leqslant c$, the *minor*

$$\{x\in S(\mathscr{D},f): v \leqslant x \leqslant c\}$$

of $S(\mathscr{D},f)$ can be understood as the translation of some integral matroid $Q(f_v^c)$ by the vector v, where the ground set rank function of $Q(f_v^c)$ satisfies for all $S\in\mathscr{B}$,

$$f_v^c(S) = \min \{f_v(T) + (c - v)(S\backslash T): T\subseteq E\}.$$

Thus, minors of submodular structures essentially are integral matroids. Choosing $v = (0, 0, \ldots, 0)$ and $c = (1, 1, \ldots, 1)$, Proposition 9.4.12 therefore implies

9.4.13. Corollary. *Let* $f: \mathscr{K} \to \mathbb{N}$ *be a submodular function on the crossing family* $\mathscr{K}$. *Then for every* $k\in\mathbb{N}$,

$$\mathbb{B}_k'(\mathscr{K},f) = \{B\subseteq E: |B| = k, |B\cap A| \leqslant f(A) \quad \text{for all} \quad A\in\mathscr{K}\}$$

is the collection of bases of a matroid on E. $\qquad\qquad\square$

9.5. Submodular Flows

A graph-theoretical model was suggested by Edmonds and Giles (1977) which generalizes the model of network flows (cf. Ford and Fulkerson 1962) and provides a unified setting for many combinatorial optimization problems.

We consider a network $N = (V, E; \underline{c}, \bar{c})$, where V is the set of nodes of a directed graph with set E of arcs, a lower *capacity function* $\underline{c}: E \to \mathbb{Z}$, and an *upper capacity* function $\bar{c}: E \to \mathbb{Z}$ such that $\underline{c} \leqslant \bar{c}$. (If $\underline{c} = 0$, we will also just use the notation $N(V, E; c)$ and refer to c as 'the' capacity function of N).

For every $e \in E$, we denote by $\partial^+ e\,(\partial^- e)$ the initial (terminal) node of the arc e. Similarly, for every $v \in V$, we set

$$\delta^+ v = \{e \in E : \partial^+ e = v\},$$
$$\delta^- v = \{e \in E : \partial^- e = v\}.$$

The *boundary* $\partial x : V \to \mathbb{Z}$ of the vector $x \in \mathbb{Z}^E$ is given by

$$\partial x(v) = \sum_{e \in \delta^+ v} x(e) - \sum_{e \in \delta^- v} x(e)$$

and extends to a function on all subsets U of nodes *via*

$$\partial x(U) = \sum_{v \in U} \partial x(v).$$

Let $\mathcal{K}$ be a crossing family on the set V of nodes of the network N and $f : \mathcal{K} \to \mathbb{Z}$ a submodular function. We seek a *feasible flow* $x : E \to \mathbb{Z}$ on N with respect to $(\mathcal{K}, f)$, i.e., a vector x satisfying

$$\underline{c} \leqslant x \leqslant \bar{c}, \tag{9.38}$$

$$\partial x(U) \leqslant f(U) \quad \text{for all} \quad U \in \mathcal{K}. \tag{9.39}$$

Thus, feasible flows are subject to the requirement that the lower and upper capacity bounds have to be observed and that the net flow out of certain given sets of nodes is limited by a submodular function.

The classical network flow models are special cases in the above setting. Taking

$$\mathcal{K} = \{\{v\} : v \in V\}, \tag{9.40}$$

$$f(v) = 0 \quad \text{for all} \quad v \in V, \tag{9.41}$$

the feasible flows are *feasible circulations* on N. The network model of Ford and Fulkerson (1962) distinguishes a *source* node $s \in V$ and a *sink* node $t \in V$ together with a *return arc* $e_{ts} = (t, s) \in E$ and investigates the optimization problem

$$\max x(e_{ts}), \text{ where } x \text{ is a feasible circulation on } N.$$

In this case, $x(e_{ts})$ is called the *value* of the flow x from the source s to the sink t.

Generally, given a weighting $w : E \to \mathbb{R}$, we may consider the optimization problem

$$\max w \cdot x, \text{ where } x \text{ is a feasible flow on } N \tag{9.42}$$

with respect to arbitrary submodular functions $f : \mathcal{K} \to \mathbb{Z}$ on crossing families $\mathcal{K}$ of nodes.

9.5.1. Example. Let $Q(f_1)$ and $Q(f_2)$ be two integral matroids on a common ground set E, and let $c \in \mathbb{N}^E$ be a common bounding vector for $Q(f_1)$ and $Q(f_2)$. We now construct a network $N = (V, \tilde{E}; \tilde{c})$.

The nodes of N are the elements of E and a disjoint copy E' together with an additional node s and sink node t.

The arcs of N are of the form (e, e'), (s, e) and (e', t) for all $e \in E$ and a return arc (t, s).

The lower capacity of N is zero and the upper capacity $\tilde{c}(e, e') = c(e)$ for all $e \in E$ and $\tilde{c}(x, y) = c(E)$ otherwise.

Letting $\mathcal{H}$ consist of all non-empty subsets of E together with all non-empty subsets of E' and defining $f: \mathcal{H} \to \mathbb{N}$ through f_1 with respect to E and f_2 with respect to E', it is clear that (9.42) in this setting is precisely the intersection problem for integral matroids. $\quad\square$

9.5.2. Example. Let $G = (V, E)$ be a directed graph and let

$$\mathcal{H} = \{D \subseteq V : \varnothing \neq D \neq V, \text{ no arc of } G \text{ leaves } D\}$$

be the crossing family of those sets of nodes which correspond to directed cuts of G. Taking

$$\bar{c} = (0, 0, \dots, 0),$$
$$\underline{c} = (-1, -1, \dots, -1),$$
$$f(U) = 1 \text{ for all } U \in \mathcal{H},$$

every feasible flow on G arises from a set of arcs which meets every directed cut. $\quad\square$

Let us now return to a network analysis from a matroidal viewpoint. Given $N = (V, E; c)$, we define for every $U \subseteq V$,

$$\hat{c}(U) = c(\hat{U}), \tag{9.43}$$

where $\hat{U} = \{e \in E; \partial^+ e \in U, \partial^- e \notin U\}$ is the *cut* of N determined by U.

9.5.3. Theorem. *Let $N = (V, E; c)$ be a network with non-negative capacity c. Let $y \in V^{\mathbb{Z}}$ be arbitrary. Then there exists a flow vector $\phi \in \mathbb{N}^E$ such that*

$$0 \leqslant \phi \leqslant c \quad and \quad \partial\phi = y$$

if and only if

$$y(V) = 0 \quad and \quad y(U) \leqslant \hat{c}(U) \quad for \; all \quad U \subseteq V. \tag{9.44}$$

Proof. Condition (9.44) is obviously necessary. We show sufficiency. Observing that the zero vector $y_0 = 0$ is the boundary vector of the (admissible) zero flow in N, it suffices to show that y_0 can be transformed to y by a sequence of feasible exchange operations. Thus, the theorem will be proved if we can verify the following claim.

Let $x = \partial \phi$ for some admissible flow ϕ in N and let s, t be two distinct nodes such that condition (9.44) holds for x', where

$$x'(v) = \begin{cases} x(v) & \text{if } v \in V \setminus \{s, t\}, \\ x(s) + 1 & \text{if } v = s, \\ x(t) - 1 & \text{if } v = t. \end{cases}$$

Then there is an admissible flow ϕ' in N with the property $\partial \phi' = x'$.

To verify the statement, we claim for every $U \subseteq V$ containing the 'source' s but not containing the 'sink' t,

$$x(U) < \hat{c}(U).$$

Indeed, if $\mathbb{B}(\hat{c})$ denotes the basis structure of the submodular system $(\mathcal{B}, \hat{c})$, $x' \in \mathbb{B}(\hat{c})$ implies

$$x(U) < x'(U) \leqslant \hat{c}(U).$$

Hence the augmenting path technique of König (1936) or Ford and Fulkerson (1962) can be applied to obtain the desired admissible flow ϕ'.

Here we call a path P from s to a node $v \in V$ in N *augmenting* with respect to ϕ if for every arc e in P,

$$\begin{cases} \phi(e) < c(e) & \text{if } e \text{ is a forward arc in } P, \\ 0 < \phi(e) & \text{if } e \text{ is a backward arc in } P. \end{cases}$$

Letting U_s consist of all vertices of N which can be reached by an augmenting path from s, we observe $t \in U_s$ since, otherwise, $t \notin U_s$ apparently implies

$$x(U_s) = \hat{c}(U_s),$$

in contradiction to the claim before.

Thus the flow ϕ may be transformed into an admissible flow ϕ' by increasing ϕ by one unit on the forward arcs and decreasing ϕ by one unit on the backward arcs along an augmenting path from s to t. The resulting flow ϕ' satisfies $x' = \partial \phi'$. $\square$

Let us come back to the network flow model (9.40) and (9.41) and call $\phi \in \mathbb{N}^E$ a *proper k-flow* from source s to sink t in the network $N = (V, E; c)$ if

$$0 \leqslant \phi \leqslant c, \tag{9.45}$$

$$\partial \phi(v) = \begin{cases} 0 & \text{if } v \in V \setminus \{s, t\}, \\ k & \text{if } v = s, \\ -k & \text{if } v = t. \end{cases} \tag{9.46}$$

Then the 'max-flow-min-cut' theorem of Ford and Fulkerson, which equates the minimum cut capacity with the maximal value of a proper flow, can be stated as

9.5.4. Corollary. *The network $N = (V, E:c)$ admits a proper k-flow from s to t if and only if*

$$k \leqslant \hat{c}(U) \qquad (9.47)$$

for all $U \subseteq V$ such that $s \in U$ and $t \notin U$.

Proof. Define $y \in \mathbb{Z}^V$ by

$$y(v) = \begin{cases} k & \text{if } v = s, \\ -k & \text{if } v = t, \\ 0 & \text{otherwise.} \end{cases}$$

Then (9.47) is equivalent to (9.44). $\qquad \square$

In the presence of two capacities in the network $N = (V, E; \underline{c} \ \bar{c})$, we set

$$c = \bar{c} - \underline{c}$$

and compute $\hat{c}$ with respect to c as in (9.43). The translation of the basis structure $\mathbb{B}(\hat{c})$ by the vector $\partial \underline{c}$ yields the basis structure $\mathbb{B}(\hat{c} + \partial \underline{c})$, which is the set of boundaries of admissible flows in N.

Hence the question whether $0 \in \mathbb{B}(\hat{c} + \partial \underline{c})$ or, equivalently, $-\partial \underline{c} \in \mathbb{B}(\hat{c})$ yields, by (9.44), Hoffman's (1960) criterion for the existence of feasible circulations:

9.5.5. Corollary. *There exists a feasible circulation in the network $N = (V, E; \underline{c}, \bar{c})$ if and only if for all $U \subseteq V$,*

$$\sum \{\underline{c}(e): \partial^+ e \in V \setminus U, \partial^- e \in U\} \leqslant \sum \{\bar{c}(e): \partial^+ e \in V \setminus U, \partial^- e \in V \setminus U\}. \qquad \square$$

We have studied network flows so far *via* the matroidal structures induced by the boundaries on the set of nodes of the network. A construction of Frank (1981), on the other hand, reveals the Edmonds–Giles model as a projection of the intersection of two basis structures on the set of arcs.

As in Example 9.5.1, we construct a bipartite graph taking E and a disjoint copy E' as set of nodes and arcs of the form (e, e'), $e \in E$. Thus each 'old' arc of the network $N = (V, E; \underline{c}, \bar{c})$ is represented as an independent arc in the auxiliary graph $G(N)$, where G has a set S, $|S| = 2 \cdot |E|$, of new nodes.

With each node $v \in V$, we associate the subset

$$\psi(v) = \{e \in E: v = \partial^+ e\} \cup \{e' \in E: v = \partial^- e\}$$

of the node set S and set for all $U \subseteq V$,

$$\psi(U) = \bigcup_{v \in U} \psi(v).$$

The function ψ maps any crossing family $\mathcal{H}$ on V onto a crossing family $\psi(\mathcal{H})$ on S.

Thus we obtain the two basis structures

$$\mathbb{B}_1 = \{y \in \mathbb{Z}^S : y(S) = 0, y(\psi(U)) \leqslant (\hat{c} + \partial\underline{c})(U), U \subseteq V\} \qquad (9.48)$$

[i.e., $\mathbb{B}_1$ is the basis structure induced by $\mathbb{B}(\hat{c} + \partial\underline{c})$] and

$$\mathbb{B}_2 = \{y \in \mathbb{Z}^S : y(S) = 0, y(\psi(U)) \leqslant f(U), U \in \mathscr{K}\} \qquad (9.49)$$

[i.e., $\mathbb{B}_2$ is induced by the basis structure $\mathbb{B}(\mathscr{K}, f; f_v(v) = 0)$]. Consider now the projection map $t : \mathbb{Z}^S \to \mathbb{Z}^T$, where $T = E$ is considered as the set of 'tails' of the arcs of G, given by

$$(ty)(e) = y(e)$$

for all $e \in T$ and $y \in \mathbb{Z}^S$.

9.5.6. Proposition. *The feasible flows $x \in \mathbb{Z}^E$ in N with respect to $(\mathscr{K}, f)$ are in one-to-one correspondence with the vectors of the projection $t(\mathbb{B}_1 \cap \mathbb{B}_2)$.*

Proof. Assume $y \in \mathbb{B}_1 \cap \mathbb{B}_2$. Then we define the vector $x = x_y \in \mathbb{Z}^E$ via, for all $e \in E$,

$$x_y(e) = y(e).$$

In view of (9.48) and (9.49), x_y satisfies the conditions (9.38) and (9.39).

Conversely, let $x \in \mathbb{Z}^E$ satisfy (9.38) and (9.39) and define $y = y_x \in \mathbb{Z}^T$ via, for all $e \in T$,

$$y_x(e) = x(e).$$

y_x is the projection of the vector $z \in \mathbb{Z}^S$, where

$$z(s) = \begin{cases} x(e) & \text{if } s = e, \\ -x(e) & \text{if } s = e'. \end{cases}$$

Since x satisfies (9.38) and (9.39), z must belong to both $\mathbb{B}_1$ and $\mathbb{B}_2$. $\square$

Proposition 9.5.6 reduces also the general submodular flow optimization problem to a matroidal intersection problem. This could in theory be solved with the matroid intersection algorithm of Section 9.2. The Dilworth embedding, however, may have to be defined on a ground set of exponential size with respect to the input size of the original problem. Hence, this algorithm could not be expected to be efficient.

The crucial step here is the passage from the matroid intersection problem in Section 9.2 to the integral matroid intersection problem in Section 9.3. Yet, taking a closer look, the situation is not as bad as it seems. With the help of Proposition 9.3.5, the intersection algorithm with respect to the Dilworth completions can be translated directly into an intersection algorithm for the integral matroids without explicit construction of the Dilworth completion.

The only additional ingredient needed is an efficient way to compute the vector rank of an integral vector with respect to an integral matroid. This amounts to minimizing a submodular function.

Many direct algorithms for submodular flow problems have been developed generalizing augmenting path techniques as well as simplex methods (e.g., Fujishige 1978b, Schönsleben 1980, Lawler and Martel 1982a, b, Frank 1984, Cunningham and Frank 1985, Barahona and Cunningham 1984). All of these algorithms assume the availability of an efficient subroutine for minimizing submodular functions.

While the construction of such a subroutine is not hard for special classes of submodular functions, it generally is a challenging problem. The results of Grötschel, Lovász, and Schrijver (1981) show that, *via* the ellipsoid method, the problem of optimizing over a given polyhedron is polynomially equivalent to the problem of testing arbitrary vectors for membership in the polyhedron. The optimization problem over integral matroid polyhedra is easily solved by the greedy algorithm. Hence, with the ellipsoid method, also the membership problem is polynomially solvable. The latter, however, consists exactly in determining the vector rank of an arbitrary vector with respect to some integral matroid.

Since the ellipsoid method cannot be considered practically efficient, the issue is to devise an efficient combinatorial procedure which solves the optimization problem

$$\min\{f(X):X \subseteq E\},$$

where $f:2^E \to \mathbb{Z}$ is a submodular function.

Submodular relaxations of the standard network flow model have also been studied from different view points. For example, the *polymatroidal* network flow model of Hassin (1981) and Lawler and Martel (1982b) considers a directed graph $N = (V, E)$ where, for every $v \in V$, two ground set rank functions F_v^+ and F_v^- with respect to ∂_v^+ and ∂_v^- are given. A *feasible circulation* $x: E \to \mathbb{N}$ now has to satisfy

$$x(A) \leqslant f_v^+(A) \quad \text{for all} \quad A \subseteq \partial_v^+, \quad v \in V, \qquad (9.50)$$
$$x(B) \leqslant f_v^-(B) \quad \text{for all} \quad B \subseteq \partial_v^-, \quad v \in V, \qquad (9.51)$$
$$x(\partial_v^+) = x(\partial_v^-) \quad \text{for all} \quad v \in V. \qquad (9.52)$$

Extending the construction of Example 9.5.1, we can formulate the polymatroidal circulation problem on N with the restrictions (9.50)–(9.52) as a submodular circulation problem on an auxiliary graph $G(N)$ of the form (9.38) and (9.39), where the set S of nodes of $G(N)$ consists of the old node set V together with two disjoint copies V^+ and V^- of V and where the members of the arc set E' of $G(N)$ are of the form $(v^+, w^-), (v, w^+),$ and (v^-, v) in accordance with the following cases:

(i) $(v, w) \in E$ implies (v^+, w^-) and $(v, w^+) \in E'$,

(ii) $v = \partial^- e$ for some $e \in E$ implies $(v^-, v) \in E'$.

We define a crossing family $\mathcal{K}$ on S consisting of

(a) the singleton sets $\{v\}$, where $v \in V$,

(b) the subsets $A^+ \subseteq V^+$, where $A \subseteq \partial^+ v$, $v \in V$,

(c) the subsets $B^- \subseteq V^-$, where $B \subseteq \partial^- v$, $v \in V$.

Choosing $f: \mathcal{K} \to \mathbb{N}$ to agree with the functions f_v^+ and f_v^- in the obvious way and letting $f(\{v\}) = 0$, the restrictions (9.50)–(9.52) take on the form (9.38) and (9.39).

Another far-reaching approach to a theory of optimization under submodular constraints was originated by Johnson (1975) and further developed by Hoffman (1976), Hoffman and Schwartz (1978), and Gröflin and Hoffman (1982) through the notion of *lattice polyhedra*. The idea thereby is to consider polyhedra similar to matroid polyhedra. The restrictions here arise from functions that are not necessarily submodular with respect to the lattice structure of the underlying power set but with respect to suitable lattice structures externally imposed onto the power set. In this setting, results analogous to Theorem 9.2.8 can be obtained. A general algorithmic theory for lattice polyhedra, however, is currently not available and it is an open question whether the Edmonds–Giles model also subsumes lattice polyhedra.

Exercises

9.1. Let $\mathcal{F} \subseteq 2^E$ be the collection of independent set of a matroid on E and let $c: E \to \mathbb{R}$ be arbitrary. Show that if $B \in \mathcal{F}$ is in accordance with the greedy algorithm, then $c(B) \geqslant c(A)$ whenever $A \in \mathcal{F}$ and $|A| = |B|$.

9.2. Give a direct (combinatorial) proof of Theorem 9.1.2.

9.3. Prove Lemma 9.1.6.

9.4. Show that every independent system can be obtained as an intersection of suitable matroid independence systems.

9.5. Show that the lower bound in Proposition 9.1.6. can, in general, not be improved (Korte and Hausmann 1978).

9.6. Show that the weighted intersection algorithm requires only a polynomially bounded number of steps with respect to $|E| = n$ provided matroid independence can be checked efficiently and addition, subtraction, and comparison of two real numbers are counted as one step each (Frank 1981).

9.7. A *branching* of the directed graph $G = (V, E)$ *rooted at* $v_0 \in V$ is a circuit-free set B of arcs of G such that for every $v \in V$, there is a unique directed path from v_0 to v using only arcs of B. Show how optimal weighted branchings can be found with the matroid intersection algorithm (Edmonds 1970).

9.8. Prove Corollary 9.2.13.

9.9. Give a matroid-theoretic proof of Corollary 9.3.9.

9.10. Show that the c-dual of a ground set rank function is a ground set rank function.

9.11. Let f be a ground set rank function and $c \in \mathbb{N}^E$ a vector and define the *convolution* $f * c$ via

$$(f * c)(S) = \min \{ f(X) + c(S \setminus X) : X \subseteq E \}.$$

Show that $f * c$ is a ground set rank function and $Q(f * c)$ is the restriction of $Q(f)$ to the set $D(c) = \{ x \in \mathbb{N}^E : x \leqslant c \}$.

9.12. Prove the *symmetric basis exchange property* of matroids (Brylawski 1973, Greene 1973): for every pair X, Y of bases of the matroid M on E and partition $X = X_1 \cup X_2$ into independent sets X_1 and X_2, there is a partition $Y = Y_1 \cup Y_2$ of Y into independent subsets Y_1 and Y_2 such that both $X_1 \cup X_2$ and $Y_1 \cup Y_2$ are bases of M. [Hint: show that y is in the sum of the contraction matroids M/X_1 and M/X_2 (cf. Woodall 1974).]

9.13. Prove Corollary 9.4.3.

9.14. Show that the optimization problem admits a finite optimal objective value if and only if the weight w is $\mathcal{D}$-compatible.

9.15. Establish the converse of Proposition 9.4.6.

9.16. Prove Lemma 9.4.9.

References

Barahona, F. and Cunningham, W.H. (1984). A submodular network simplex method. *Math. Progr. Study* **22**, 9–31.

Barthélemy, J.P., Leclerc, B. and Monjardet, B. (1984). Ensembles ordonnés et taxomnomie mathématique. *Ann. Disc. Math.* **23**, 523–48.

Bixby, R.E. (1982). Matroids and operations research, in: *Advanced Techniques in the Practice of Operations Research* (H.J. Greenberg *et al.*, eds.), North Holland, Amsterdam, 333–458.

Bland, R.G. (1977). A combinatorial abstraction of linear programming. *J. Comb. Theory Ser. B* **23**, 33–57.

Boruvka, O. (1926). On jistem problemu minimalnim. *Prace Moravske Prirodovedecke Spolecnosti* **3**, 37–53.

Brylawski, T.H. (1973). Some properties of basic families of subsets. *Discr. Math.* **6**, 333–41.

Chvátal, V. (1983). *Linear Programming*. Freeman, New York–San Francisco.

Crawley, P. and Dilworth, R.P. (1973). *Algebraic Theory of Lattices*. Prentice-Hall, Englewood Cliffs, New York.

Cunningham, W.H. and Frank, A. (1985). A primal-dual algorithm for submodular flows. *Math. of Oper. Res.* **10**, 251–62.

Derigs, U., Goecke, O. and Schrader R. (1984). Bisimplicial edges, Gaussian elimination, and matchings in bipartite graphs, in *Graphtheoretic Concepts in Computer Science* (U. Pape, ed.), Trauner-Verlag, Linz, 79–87.

Dunstan, F.D.J., Ingleton, A.W. and Welsh, D.J.A. (1972). Supermatroids, in *Proc. Conf. Comb. Math., Math. Inst., Oxford*, 72–122.

Edmonds, J. (1965). Paths, trees and flowers. *Canad. J. Math.* **14**, 449–67.

Edmonds, J. (1968). Matroid partition. *Math. of the Decision Sci., Amer. Math. Soc. Lectures in Appl. Math.* **11**, 335–45.

Edmonds, J. (1970). Submodular functions, matroids, and certain polyhedra, in *Combinatorial Structures and their Applications* (R. Guy *et al.*, eds.), Gordon and Breach, New York, 69–87.

Edmonds, J. (1971). Matroids and the greedy algorithm. *Math. Progr.* **1**, 127–36.

Edmonds, J. (1979). Matroid intersection. *Ann. Discr. Math.* **4**, 39–49.

Edmonds, J. and Fulkerson, D.R. (1965). Transversals and matroid partition. *J. Res. Nat. Bur. Stand.* **69B**, 147–53.

Edmonds, J. and Giles, R. (1977). A min–max relation for submodular functions on graphs. *Ann. Discr. Math.* **1**, 185–204.

Edmonds, J. and Giles, R. (1984). Total dual integrality of linear inequality systems, in *Progress in Combinatorial Optimization* (W.R. Pulleyblank, ed.), Academic Press, 117–129.

Faigle, U. (1979) The greedy algorithm for partially ordered sets. *Discr. Math.* **28**, 153–9.

Faigle, U. (1980). Geometries on partially ordered sets. *J. Comb. Theory Ser. B* **28**, 26–51.

Faigle, U. (1985). On ordered languages and the optimization of linear functions by greedy algorithms. *J. ACM* **32**, 861–70.

Ford, L.R. and Fulkerson, D.R. (1962). *Flows in Networks.* Princeton University Press, Princeton, New York.

Frank, A. (1981). A weighted matroid intersection algorithm. *J. Algo.* **2**, 328–36.

Frank, A. (1982). An algorithm for submodular functions on graphs. *Ann. Discr. Math.* **16**, 97–120.

Frank, A. (1984). Finding feasible vectors in Edmonds–Giles polyhedra. *J. Comb. Theory Ser. B* **36**, 221–39.

Frank, A. and Tardos, E. (1982). Matroids from crossing families. *Report No. 82210-OR, Inst. für Operations Research, Universität Bonn* (to appear in *Proceedings Sixth Hungarian Combinatorial Colloquium, Eger, 1981*).

Fujishige, S. (1978a). Polymatroid dependence structure of a set of random variables. *Inform. and Control* **39**, 55–72.

Fujishige, S. (1978b). Algorithms for solving the independent flow problem. *J. Oper. Res. Soc. Japan* **21**, 189–203.

Fujishige, S. (1984a). Submodular systems and related topics. *Math. Progr. Study* **22**, 113–31.

Fujishige, S. (1984b). Theory of submodular programs: A Fenchel-type min–max theorem and subgradients of submodular functions. *Math. Prog.* **29**, 142–55.

Fujishige, S. (1984c). Structures of polyhedra determined by submodular functions. *Math. Progr.* **29**, 125–41.

Garey, M.R. and Johnson, D.S. (1979). *Computers and Intractability – A Guide to the Theory of NP-Completeness.* W.H. Freeman, San Francisco.

Gröflin, H. and Hoffman, A.J. (1982). Lattice polyhedra II: generalizations, constructions and examples. *Ann. Discr. Math.* **15**, 189–203.

Greene, C. (1973). A multiple exchange property for bases. *Proc. Amer. Math. Soc.* **39**, 45–50.

Grötschel, M., Lovász, L. and Schrijver, A. (1981). The ellipsoid method and its consequences in combinatorial optimization. *Combinatorica* **1**, 169–97.

Hassin, R. (1981). On network flows. *Networks* **12**, 1–21.

Helgason, T. (1974). Aspects of the theory of hypermatroids, in *Hypergraph Seminar, Springer Lecture Notes* **411**, 191–214.

Hoffman, A.J. (1960). Some recent applications of the theory of linear inequalities to extremal combinatorial analysis, in *Combinatorial Analysis* (R.E. Bellman and M. Hall Jr., eds.), Amer. Math Soc., Providence, R.I., 113–27.

Hoffman, A.J. (1974). A generalization of max-flow min-cut. *Math. Progr.* **6**, 352–49.

Hoffman, A.J. (1976). On lattice polyhedra III: blockers and anti-blockers of lattice clutters. Math. Progr. Study **8**, 197–207.

Hoffman, A.J. and Schwartz, D.E. (1978). On lattice polyhedra, in *Combinatorics* (A. Hajnal and V.T. Sós, eds.), Bolyai I. Math. Soc., North Holland, Amsterdam, 593–8.

Iri, M. (1983). Applications of matroid theory, in *Mathematical Programming – The State of the Art* (A. Bachem *et al.*, eds.), Springer-Verlag, Heidelberg, 158–201.

Iri, M. and Tomizawa, N. (1975). An algorithm for finding an optimal 'independent assignment'. *J. Oper. Res. Soc. Japan* **58A**, 33–40.

Jenkyns, H.A. (1976). The efficacy of the 'greedy' algorithm. in *Proc. 7th S.-E. Conf. Combinatorics, Graph Theory, and Computing,* 341–50.

Jensen, P.M. and Korte, B. (1982). Complexity of matroid property algorithms *SIAM J. Comp.* **11**, 184–190.

Johnson, E. (1975). On cut set integer polyhedra. *Cahiers du Centre de Recherche Opérationelle*, **17**, 235–51.

König, D. (1936). *Theorie der endlichen und unendlichen Graphen.* Leipzig, 1936; Chelsea reprint, New York, 1950.

Korte, B. and Hausmann, D. (1978). An analysis of the greedy heuristic for independence systems. *Ann. Discr. Math.* **2**, 65–74.

Kruskal, J.B. (1956). On the shortest spanning subtree of a graph and the travelling salesman problem. *Proc. Amer. Math. Soc.* **7**, 48–50.

Lawler, E.L. (1975). Matroid intersection algorithms. *Math. Progr.* **9**, 31–56.

Lawler, E.L. (1976). *Combinatorial Optimization: Networks and Matroids.* Holt, Rinehart, and Winston, New York.

Lawler, E.L. and Martel, C.U. (1982a). Computing maximal 'polymatroidal' network flows. *Math. of Oper. Res.* **7**, 334–47.

Lawler, E.L. and Martel, C.U. (1982b). Flow network formulations of polymatroid optimization problems. *Ann. Discr. Math.* **16**, 189–200.

Lovász, L. (1980). Matroid matching and some applications. *J. Comb. Theory Ser. B* **28**, 208–36.

Lovász, L. (1981). The matroid matching problem, in *Algebraic Methods in Graph Theory*, vol. II (L. Lovász and V.T. Sós, eds.), *Coll. Math. Soc. I. Bolyai* **25**, North-Holland, Amsterdam, 495–517.

Lovász, L. (1983). Submodular functions and convexity, in *Mathematical Programming – The State of the Art* (A. Bachem, M. Grötschel, and B. Korte, eds.), pp. 235–57. Springer-Verlag. Berlin–Heidelberg–New York–Tokyo.

McDiarmid, C.J.M. (1975). Rado's theorem for polymatroids. *Math. Proc. Camb. Phil. Soc.* **78**, 263–81.

Minty, G.J. (1966). On the axiomatic foundations of the theories of directed linear graphs, electrical networks and network-programming. *J. Math. and Mech.* **15**, 485–520.

Monge, G. (1781). Deblai et remblai, Mémoires de l'Académie de Sciences, Paris.

Nash-Williams, C.St.J.A. (1961). Edge-disjoint spanning trees of finite graphs. *J. Lond. Math. Soc.* **36**, 445–50.

Nash-Williams, C.St.J.A. (1964). Decomposition of finite graphs into forests. *J. Lond. Math. Soc.* **39**, 12.

Papadimitriou, C.H. and Steiglitz, K. (1982). *Combinatorial Optimization – Algorithms and Complexity.* Prentice-Hall, Englewood Cliffs, NJ.

Rado, R. (1957). Note on independence functions. *Proc. Lond. Math. Soc.* **7**, 300–20.

Rockafellar, R.T. (1969). The elementary vectors of a subspace of $\mathbb{R}^N$, in *Combinatorial Mathematics and its Applications* (R.C. Bose and T.A. Dowling, eds.), UNC Press, Chapel Hill, NC, 104–27.

Rosenmüller, J. (1983). Nondegeneracy problems in cooperative game theory, in *Mathematical Programming – The State of the Art* (A. Bachem et al., eds.). Springer-Verlag, Heidelberg, 391–416.

Schönsleben, P. (1980). Ganzzahlige Polymatroid-Intersektions-Algorithmen. Ph.D. Thesis, ETH Zürich.

Schrijver, A. (1982). Submodular functions. *Note AE N5/82*, Faculty of Actuarial Science, University of Amsterdam.

Schrijver, A. (1983). Min-max results in combinatorial optimization, in *Mathematical Programming – The State of the Art* (A. Bachem et al., eds.). Springer-Verlag, Heidelberg, 439–500.

Schrijver, A. (1984). Total dual integrality from directed graphs, crossing families, and sub- and supermodular functions, in *Progress in Combinatorial Optimization* (W.R. Pulleyblank, ed.), Academic Press, Toronto, 315–361.

Seymour, P.D. (1977). The matroids with the max-flow min-cut property. *J. Comb. Theory Ser. B* **23**, 189–222.

Seymour, P.D. (1980). Decomposition of regular matroids. *J. Comb. Th. B* **28**, 305–60.

Tardos, E. (1983). Generalized matroids and supermodular colorings. *Report AE 19/83*, University of Amsterdam.

Tong, Po, Lawler, E.L., and Vazirani, V.V. (1984). Solving the weighted parity problem for gammoids by reduction to graphic matching, in *Progress in Combinatorial Optimization* (W.R. Pulleyblank, ed.), Academic Press, New York, 363–74.

Tutte, W.T. (1961). On the problem of decomposing a graph into *n* connected factors. *J. London Math. Soc.* **36**, 221–30.

Welsh, D.J.A. (1976). *Matroid Theory*. Academic Press, London.

Welsh, D.J.A. (1982). Matroids and combinatorial optimization, in *Matroid Theory and its Applications* (A. Barlotti, ed.), Liguori Editore, Napoli, 323–416.

Whitney, H. (1935). On the abstract properties of linear dependence. *Amer. J. Math.* **57**, 509–33.

Woodall, D.R. (1974). An exchange theorem for bases of matroids. *J. Comb. Theory* **16**, 227–8.

INDEX